T0350990

Computer Arithmetic in Practice

Computer Arithmetic in Practice: Exercises and Programming is a simple, brief introductory volume for undergraduate and graduate students at university courses interested in understanding the foundation of computers. It is focused on numeric data formats and capabilities of computers to perform basic arithmetic operations. It discusses mainly such topics as:

- Basic concepts of computer architecture
- Assembly language programming skills
- Data formats used to express integer and real numbers
- Algorithms of basic arithmetic operations
- Short overview of nonlinear functions evaluation
- Discussion on limited number representation and computer arithmetic
- Exercises and programming tasks

This book provides an accessible overview of common data formats used to write numbers in programming languages and how the computer performs four basic arithmetic operations from the point of view of the processor instruction set. The book is primarily didactic in nature, therefore the theoretical information is enriched with many numerical examples and exercises to be solved using a 'sheet of paper and a pencil'. Answers are provided for most of the tasks.

The theoretical discussed issues are illustrated by listings of algorithms presenting the way to implement arithmetic operations in low-level language. It allows development the skills of optimal programming, taking into consideration the computer architecture and limitations. Creating software using low-level language programming, despite the initial difficulties, gives the ability to control the code and create efficient applications. This allows for effective consolidation of knowledge and acquisition of practical skills required at this stage of education, mainly a specialist in the field of information technology, electronics, telecommunications, other related disciplines, or at the level of general education with introduction to information technology. It may be also useful for engineers interested in their own professional development and teachers as well.

 **Sławomir Gryś** is a university professor at Częstochowa University of Technology, Poland. He has conducted many courses focused on analog electronics, logical devices, foundations of computer architecture and organization, low-level programming techniques in assembly and C languages for embedded systems, image processing and recognition. He is the author or co-author of several scientific monographs, book chapters, academic textbooks, patents and more than 60 papers in journals and domestic and international conference proceedings in Poland, Germany, Canada and Australia.

Computer Arithmetic in Practice
Exercises and Programming

Sławomir Gryś

CRC Press
Taylor & Francis Group
Boca Raton London New York

CRC Press is an imprint of the
Taylor & Francis Group, an **informa** business

Designed cover image: Shutterstock_587427896

First edition published 2024
by CRC Press
6000 Broken Sound Parkway NW, Suite 300, Boca Raton, FL 33487-2742

and by CRC Press
4 Park Square, Milton Park, Abingdon, Oxon, OX14 4RN

CRC Press is an imprint of Taylor & Francis Group, LLC

Library of Congress Cataloging-in-Publication Data
Names: Gryś, Sławomir, author.
Title: Computer arithmetic in practice : exercises and programming / SŁŁawomir Gry
Other titles: Arytmetyka komputerâow. English
Description: First edit | 1 Boca Raton, FL : CRC Press, [2 | Translation of: Arytmetyka komputerał |
Includes bibliographical references and in |
Identifiers: LCCN 2023010741 | ISBN 9781032425634 (hbk) | ISBN 9781032425658 (pbk) | ISBN 9781003363286 (ebk)
Subjects:
Classification: LCC QA76.9.C62 G7913 2024 | DDC 004.01/51--dc23/eng/20230404
LC record available at https://lccn.loc.gov/2023010741

ISBN: 978-1-032-42563-4 (hbk)
ISBN: 978-1-032-42565-8 (pbk)
ISBN: 978-1-003-36328-6 (ebk)

DOI: 10.1201/9781003363286

Typeset in Sabon
by MPS Limited, Dehradun

Dedicated to my family
(Thanks for forbearance and patience)

Contents

Preface

Almost 15 years have passed since the first edition of this book was published in Poland by Polish Scientific Publisher PWN with ISBN 9788301151317. The book was positively rated by the academic community, both students and teachers including students of several universities, on which I had the satisfaction with teaching them computer arithmetic as one of the main topics related to foundations of the computer architecture. The measure of book popularity may be, in my opinion, listing it as 'further readings' in the syllabus of many university courses related to the computer science as a field of education taught in Poland. I really hope that the first edition of book fulfilled its role, which encouraged me to start preparation of work on the revised and extended version of a book for a wider international audience. This textbook was not aimed to compete with classical books, those provide the complete knowledge in this topic and are well written, but can be good choice as a first look at topic. The book is rather a simple, brief introductory volume for undergraduate and graduate students at university courses related to the introduction to computer science. It may be also useful for design engineers interested in their own professional development.

The two topics are highlighted in this textbook: explanation how the computers realize some relatively simple arithmetic operations for numbers stored in various formats using simple method of 'paper and pencil' and its realization in low-level programming language considering the features and limitations of the instruction list of real microprocessor. The simplest possible architecture was chosen to facilitate the understanding the code created for real microprocessor. The assembly code presented in this book can be freely downloaded from the website (https://routledgetextbooks. com/textbooks/instructor_downloads/). The book is primarily didactic in its nature, and therefore the presented required theoretical information has been illustrated with numerous examples and exercises both in calculation, algorithms and coding in assembly language aimed for self-assessment. Many examples are giving the occasion for understanding the link between theory and practice and expand student's knowledge and skills. The solutions to the exercises are included in Appendix C, except those marked with an asterisk character.

This book, compared with the first Polish edition, has been enriched with, among other items, the theoretical basis and discussion of selected algorithms, the recommendation of the world-wide accepted IEEE P-754 standard with its all revisions. Completely new topic concerns selected methods of computing nonlinear functions. The discussed content was also illustrated with more examples. The original text was revised and corrected. An update of the content referring to the current state of the art was essential improvement. The author has made many efforts to ensure that the presentation of the discussed topic is as clear and transparent as much as possible. The programs written in assembly codes have been thoroughly tested on various combinations of data, but there is no guaranty of correct operation for any data. So, they are delivered as is with no claiming option. Any comments on book are very welcome and please forward them to the one of e-mail addresses, i.e. slawomir.grys@pcz.pl (Częstochowa University of Technology) or private slavo5.sg@gmail.com.

I hope that several features make the textbook accessible for the reader, i.e. friendly presentation, numerous examples also implemented in assembly code of a real microprocessor, theory well balanced with practice, topics limited to the most typical and important for practitioners.

The book would have not been written and published without interaction with many people. The author would like to say 'many thanks' to the reviewers for their valuable comments. They would certainly contributed to improving the quality of our work, its completeness and legibility. Special thanks are also due for Editor Ms. Gabriella Williams – Information Security, Networking, Communication and Emerging Technologies from CRC Press. She was in touch with me from the moment of submitting the textbook proposal, reviewing phase and solved all technical and organizational issues related to the preparation of the manuscript for publishing. I also wish to thank my students for all discussions and comments on presented material during common work at university. Support by the Częstochowa University of Technology, particularly Faculty of Electrical Engineering, as well as excellent workplace and motivation, is acknowledged. This essential support is greatly appreciated. Finally, the invaluable understanding of my wonderful wife, Agnieszka, and children, Antonina and Aleksander, was indisputable condition to the success of this work. The scope of this book is as follows:

- Chapter 1 presents an overview of the general features and architecture of simple microprocessors: main components as ALU, registers, flags, stack and instruction set. The 1-bit logical and arithmetical operations are shown as being the base for any more complex operations. Assembly language and tools, i.e. assembler and linker as needed for obtaining machine code ready to run on microprocessor, are discussed. The way to create the right code in assembly language avoiding wrong syntax causing bugs is done as an introduction for understanding the listings given in next chapters. Furthermore, typical file formats such as BIN,

HEX and ELF are mentioned and HEX format explained on real example.

- Chapter 2 introduces the way of representation of unsigned and signed numbers in fixed point format mainly aimed for integer numbers. The fractional part is also considered as requested in some cases. The considerations are carried out both for unsigned and signed numbers. The following formats are discussed, i.e. natural binary code, hexadecimal, unpacked and packed binary coded decimal codes and ASCII. For signed numbers the sign-magnitude and 2's complements representation as practical use of complementation theory are presented. The methods of conversion from one to another format are provided. All formats are illustrated with examples of number and conversions by software implementation in assembly code.

- Chapter 3 discusses the principles of four elementary basic arithmetic operations and its realization in assembly language. Operations are performed for all formats presented in the previous chapter. Four operations, that means, addition, subtraction, multiplication and division are exclusively for BIN format as being easy to realize. First three of them are discussed also for signed numbers represented in 2's complement format and addition together with subtraction for the others, i.e. BCD, ASCII and sign-magnitude. Nonlinear function approximation methods are shortly mentioned using iterative techniques or simple lookup tables. For some cases, the missing arithmetic instructions of real microprocessor were programmatically emulated according to 'filling gaps' strategy.

- Chapter 4 deals with number representation in floating-point format for expressing the real numbers. Non-normalized numbers are introduced. The main topic is a worldwide accepted and applied IEEE 754 standard as a hardware independent. Among others, the following issues are highlighted: single and double precision, special values and exceptions. The changes imposed by IEEE 854 update and related to the need of support shorter than single and longer than double precision new formats are announced. Additionally, some key-value features of a FPU floating point unit as specialized arithmetic coprocessor were pointed out. The universal method of conversion to another radix is provided.

- Chapter 5 similarly to Chapter 3 presents the rules for four basic arithmetic operations as addition, subtraction, multiplication and division illustrated with numerical examples and exercises. The very simplified form of a floating-point numbers format was chosen as it seemed to be more accurate and readable than a format complying with IEEE standard requirements. It was applied to present the arithmetic operations implementation in assembly code. The listings are really not short but also not too hard to understand and rebuilt for practical applications. The normalization and denormalization routines needed for proper operation execution and ensuring that the output number will keep the ensured format are pointed too.

- Chapter 6 is devoted to possible errors due to limited precision of number representation. Error magnification caused by error propagation is also explained with appropriate examples. This issue is important in case of single arithmetic operation and much more for algorithms based on multi iterations or matrix operations. The problem was only signaled as essential and noteworthy. Unfortunately, no universal solution for this issue was proposed so far that could be applicable for practice. Ignoring the computer limitations or using wrong number format can cause quite freaky incorrect results.
- Appendixes are aimed to ease reading and understanding the chapters. Appendix A presents the range numbers for the assumed number of allowed bits. It can be useful in evaluating the minimal number of bits needed to express the input numbers or result of arithmetical operation. Comparison is performed for numbers with fractional parts both for unsigned and signed formats. The binary, 2's complement and sign-magnitude are considered. Appendix B is related to the preview one and is limited to formats and numbers of bits, mainly multiples of eight, commonly used in high-level languages like Delphi Pascal, C/C++, Java and Microsoft Visual Basic. The third, Appendix C, provides the solutions to almost all exercises from book chapters.

The book is attached with three appendixes useful during reading the chapters, i.e. range numbers for assumed number of allowed bits, numerical data types with ranges in some common high-level languages and solutions to almost all exercise.

Finally, I wish you a pleasant reading, and, what is the most important, a self-practice and inspiration for your own software solutions.

Sławomir Gryś (Author)
Częstochowa (Poland)
January 2023

Chapter 1

Basic Concepts of Computer Architecture

1.1 THE 1-BIT LOGICAL AND ARITHMETICAL OPERATIONS

Let's look at the dictionary definition of computer cited after:

> < Latin: computare>, electronic digital machine, an electronic device designed to process information (data) represented in digital form, controlled by a program stored in memory.

> [Encyclopedia online PWN 2023, https://encyklopedia.pwn.pl/]

> an electronic machine that is used for storing, organizing, and finding words, numbers, and pictures, for doing calculations, and for controlling other machines.

> [Cambridge Academic Content Dictionary 2023, https://dictionary.
> cambridge.org/dictionary/english/computer]

> a programmable electronic device designed to accept data, perform prescribed mathematical and logical operations at high speed, and display the results of these operations.

> [https://www.dictionary.com/browse/operation]

Unfortunately, the above definitions ignore the outstanding achievements of many pioneers of the age of mechanical calculating machines; to cite just a few names: Schickard, Pascal, Leibniz, Stern, Jacquard, Babbage, linking the emergence of the computer with the development of electro-technology in the second half of the 20th century. Those interested in the history of the evolution of computing machines are encouraged to read [Augarten 1985, Ceruzzi 1998, McCartney 1999, Mollenhoff 1988, and Pollachek 1997]. As it can be shown, all the computer functions mentioned in the definition (and thus the performance of calculations, which is the subject of this book) can be realized by limited set of logical functions and data transfers from and to computer's memory.

DOI: 10.1201/9781003363286-1

Information in a computer is expressed by set of distinguishable values, sometimes called states. In binary logic, these states are usually denoted by symbols {0,1} or {L,H}. The symbols {0,1} are applied much more often because it is associated with commonly used numbers used to express numerical values, so it will be also used further in this book. The symbols L (low) and H (high) are used in digital electronics for describing the theory of logic circuit. In physical implementations, they translate into two levels of electric voltage, e.g. 0 V and 3.3 V or current 4 mA and 20 mA (so-called current loop). In the most cases of computer architectures and communication technologies, the positive logic is used, where 1 is the distinguished state and identical to H.

The rule of operation of computer is just data processing that means convert the input data into output data according to the given algorithm. Because any data is represented as a combination of bits, i.e. 0 and 1 states, the well-known logical operations can be applied for bit manipulation.

Let's start with 1-bit logical operations:

- inversion (denoted by '/')
  $/0 = 1, /1 = 0$
- logical OR (denoted by '∪')
  $0 \cup 0 = 0, 0 \cup 1 = 1, 1 \cup 0 = 1, 1 \cup 1 = 1$
- logical AND (denoted '∩')
  $0 \cap 0 = 0, 0 \cap 1 = 0, 1 \cap 0 = 0, 1 \cap 1 = 1$
- logical XOR (denoted by '⊕')
  $0 \oplus 0 = 0, 0 \oplus 1 = 1, 1 \oplus 0 = 1, 1 \oplus 1 = 0.$

Some additional notes on logical operations are as follows:

1. Inversion can be treated as complement of value from the set of {0,1}.
2. Logical OR is equivalent to the function of alternative.
3. Logical AND is equivalent to the function of conjunction.
4. Logical XOR for more than two inputs is equivalent to non-parity function, e.g. $1 \oplus 0 \oplus 1 \oplus 1 = 1$ and $1 \oplus 1 \oplus 0 \oplus 0 = 0$.

It should be mentioned that set of two operations {/,∩} or {/,∪} is sufficient to emulate the others. This is because de Morgan's laws apply as follows:

- $/(A \cap B) = /A \cup /B$

and

- $/(A \cup B) = /(A) \cap /B$

where A, B – logical inputs 0 or 1.

According to the principles given above, the logic gates operate, being the smallest logical element for data processing realized as digital electronic circuits. Furthermore, the computer can be considered as very complex combination of logical gates with feedback loop from outputs to inputs. The feedback is needed to realize the influence of stored data on the current output results. Hence, in the theory of computation, the computer is now an example of sophisticated finite state sequential machine. This topic as being as not strictly related to the main book topic will be not continued here.

In addition to logical operations, 1-bit arithmetic operations can be defined:

- arithmetic sum (denoted '+')

  $0 + 0 = \{0,0\}$ $0 + 1 = \{0,1\}$ $1 + 0 = \{0,1\}$ $1 + 1 = \{1,0\}$

- arithmetic difference (denoted '−')

  $0 - 0 = \{0,0\}$ $0 - 1 = \{1,1\}$ $1 - 0 = \{0,1\}$ $1 - 1 = \{0,0\}$

- arithmetic sum/difference modulo 2 (denoted '⊕')

  $0 \oplus 0 = 0$ $0 \oplus 1 = 1$ $1 \oplus 0 = 1$ $1 \oplus 1 = 0$

- arithmetic multiplication (denoted '*')

  $0 * 0 = 0$ $0 * 1 = 0$ $1 * 0 = 0$ $1 * 1 = 1$

Some additional notes about arithmetic operations:

1. Arithmetic sum returns the result of an operation in the form of a pair of bits {carry, result}.
2. Arithmetic difference returns the result of an operation in the form of a pair of bits {borrow, result}.
3. Arithmetic sum/difference modulo 2 returns an identical result to the logical XOR.
4. Arithmetic multiplication of 1-bit arguments returns identical result as logical AND.
5. The rule of arithmetic multiplication is just a multiplication table for binary numbers. Its simplicity undoubtedly draws your attention!

The presented arithmetic operations can be realized by logical operations, and therefore gates, which is an advantage of the zero-one system. The described 1-bit operations are the basis for operations on multi-bit arguments. The result of a logical operation is a composite of the results of 1-bit logical operations of individual bit pairs. Let's illustrate it with an example as below.

Example 1.1: Two arguments 4-bit logical operations:

$$\begin{array}{llll} & 1100 & 1101 & 1101 \\ /\,(1001) = 0110 & \cup\ \underline{0110} & \cap\ \underline{0001} & \oplus\ \underline{1001} \\ & -\ 1110 & 0001 & 0100 \end{array}$$

The practical meaning of logical operations results from their properties. If we mark one of the arguments with A and treat the other as the so-called mask, the chosen bits can be cleared (value is 0) by a logical AND operation, set (value is 1) by a logical OR and inverted by XOR function. These conclusions result from the following observations:

$A \oplus 0 = A$ $A \oplus 1 = /A$ $A \cup 0 = A$ $A \cup 1 = 1$ $A \cap 0 = 0$ $A \cap 1 = A$

Example 1.2: Use of binary logical operations for bit manipulation:

$$\begin{array}{llll} a_3 & a_2 & a_1 & a_0 \\ \cup\ \underline{0\ \ \ 1\ \ \ 1\ \ \ 0} \\ a_3 & 1 & 1 & a_0 \end{array} \qquad \begin{array}{llll} a_3 & a_2 & a_1 & a_0 \\ \cap\ \underline{0\ \ \ 1\ \ \ 1\ \ \ 0} \\ 0 & a_2 & a_1 & 0 \end{array} \qquad \begin{array}{llll} a_3 & a_2 & a_1 & a_0 \\ \oplus\ \underline{0\ \ \ 1\ \ \ 1\ \ \ 0} \\ a_3 & /a_2 & /a_1 & a_0 \end{array}$$

In the book, the reader will find the exercises for self-assessment. The solutions to exercises are attached in Appendix C.

Exercise 1.1: Determine the result of logical operations:

$$\begin{array}{llll} a_3 & a_2 & 0 & 1 \\ \cup\ \underline{a_3\ \ 1\ \ a_1\ \ 1} \\ ? & ? & ? & ? \end{array} \qquad \begin{array}{llll} a_3 & a_2 & 0 & 0 \\ \cap\ \underline{1\ \ a_2\ \ 0\ \ a_0} \\ ? & ? & ? & ? \end{array} \qquad \begin{array}{llll} 0 & 1 & a_1 & a_0 \\ \oplus\ \underline{a_3\ \ 1\ \ 1\ \ a_0} \\ ? & ? & ? & ? \end{array}$$

Logic and arithmetic operations are not only the domain of computer science. They are used by electronics engineers for designing digital systems in PLD/FPGA programmable logic structures. The knowledge of Boolean algebra, methods of synthesis and analysis of combinatorial and sequential circuits is necessary here. This subject is discussed in books on computer architecture and organization or digital electronics. If you are interested, I refer you to generally known books, e.g. [Null 2018, Tietze 2002 and Vladutiu 2012]. For further consideration, is it enough if we will be aware that logical and arithmetic operations are performed in hardware by element of the processor called the arithmetic logic unit, abbreviated as ALU? Modern processors usually are equipped with additional resource, the floating-point unit (FPU) working with numbers in floating-point format. These topics are discussed in Chapters 4 and 5.

1.2 ARCHITECTURE OF SIMPLE MICROPROCESSOR

The aim of the chapter is to familiarize with the basic elements of the processor, which will be referred to in programs showing how to implement arithmetic operations. Figure 1.1 shows a simplified structure of a classical processor with which popular microcontroller of 8-bit 8051 family is compatible. The figure omits such elements which are not important from the point of view of the subject matter of this book. These are up/down counting timers, serial transceiver/receiver, interrupt controller, etc.

The primary reason for choosing CPU based on 8051 architecture is its simplicity, an ideal feature from a didactic point of view and objectives of this book. This core did not lose much its popularity despite many years since its release to the market. Today, 8051s are still available as integrated circuits offered by many companies and supported by integrated development environments, but they are mainly used as silicon-based intellectual property cores. These cores, available in the source code of a hardware description language (such as VHDL or Verilog) or as an FPGA network list, are typically integrated into systems embedded in products such as USB flash drives, home appliances, and wireless communication system chips. Designers use 8051 silicon IP cores due to their smaller size and lower performance compared to 32-bit processors.

The 8051 microcontrollers were developed by Intel, so it is not surprising that the syntax of its instruction list is close to that deserved family of 8086

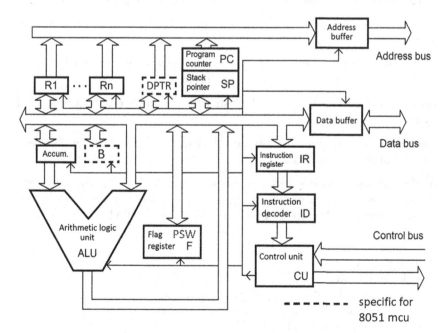

Figure 1.1 Simplified architecture of the classical processor.

processors, continued to the present day in Intel Core architecture. The differences seen from the programmer's point of interest occur mainly in word length, which is related to the width of registers, or the number and variants of instructions and resources like registers, internal memories, number of execution units, etc. A processor with a complex architecture, on the one hand, would provide more possibilities to implement various data formats, e.g. storing real or complex numbers and more advanced arithmetic operations, but probably at the cost of the clarity of the presented content.

The implementation in code of arithmetic operations presented in the book was developed just for the 8051 complying microprocessors. Those can be reused as code snippets or ready to use algorithms in the targeted user programs, after an adaptation to the architecture and list of instructions for specific targeted processor. The reader interested in deeply studying the architecture details of other processors is referred to [Baer 2010, Blaauw 1997, Hamacher 2012, Metzger 2007, Patterson 2014 and Stallings 2008].

As mentioned, the components shown in Figure 1.1 are common to most processors, so it is useful to become familiar with their functions:

Data buffer – a register that stores data exchanged between the components of the processor and external memory or an input-output device.

Address buffer – a register that stores the address of an external memory cell or input-output device.

Program memory – read only memory for storing the program code.

Data memory – read and write memory intended for data storage used by program.

Instruction register – a register that stores the instruction code fetched from the program memory (is working as a pointer).

Instruction decoder – translates the instruction to microcode being executed by the internal units of processor.

Control unit – responsible for coordination of data transfer between internal units of processor.

Arithmetic logical unit – performs basic arithmetic and logical operations on arguments stored in processor registers or memory and determines flags of status register. Operations performed by ALU (of 8051 CPU):

- logical OR,
- logical AND,
- logical XOR,
- addition (of unsigned and signed numbers),
- subtraction (of unsigned and signed numbers),
- correction after BCD addition,
- unsigned multiplication,

- unsigned division,
- comparison of two sequences of bits,
- rotations,
- clearing/setting and inverting selected bits.

 DID YOU KNOW?

There are usually additional instructions (including jumps/branches) available in 32/64-bit processors, e.g. Intel Core family:

- decimal correction after subtraction,
- decimal correction after multiplication,
- decimal correction before division,
- multiplication of signed numbers,
- division of signed numbers,
- comparison of two unsigned/signed numbers,
- shifting with or without extra bit.

Their absence in the 8051 is not a relevant problem due to possibility of in code emulation.

The 8051 flags set as a result of arithmetic operations are:

- overflow (OV or V),
- carry (C or CY),
- auxiliary carry (AC or half carry HC).

The meaning of the specified flags:

- OV – set when a range overrun occurs for signed numbers in the 2's complement notation after arithmetic addition or subtraction; also signals an attempt to divide by zero; for single-byte operations, the allowed range for numbers in the 2's complement code is <–128,127>.
- C – set when there is a carry from the 7th bit to the 8th (out of byte) after arithmetic addition or a borrow from the 8th bit to the 7th after arithmetic subtraction, signals an out-of-range result for numbers in the natural binary and packed BCD systems; also used as an extra bit during rotation instruction; for single-byte operations, the allowed range for natural binary numbers is <0,255>, and for packed BCD <0,99>.

Table 1.1 Bits of the PSW Register

PSW.7	PSW.6	PSW.5	PSW.4	PSW.3	PSW.2	PSW.1	PSW.0
C	AC	F0	RS1	RS0	OV	–	P

- AC – set when there is a carry from 3th to 4th bit (to the next nibble) after arithmetic addition or borrow from bit 4th to 3th bit after arithmetic subtraction; signals the need to perform correction of result for numbers in the packed BCD notation.

For 8051-compatible microcontrollers, the flags are stored in the PSW register presented in Table 1.1.

The flag can be tested by conditional jump/branch instructions or taken as a third input argument by arithmetic operations. The exact relationship between the flag and the instruction is presented in the next subchapter discussing the 8051-microcontroller instruction list (ISA).

The meanings of the rest of PSW register flags are as follows:

- P – determines the parity of the number of ones in the accumulator, $P = 1$ if this is odd and $P = 0$ if even.
- F0, PSW.1 – flags of general use.
- RS1, RS0 – register bank selection flags R0...R7.

The flags F0, PSW.1 can be used for any purpose, e.g. storing sign bits of numbers; they can be tested by conditional jump instructions.

The flags RS1 and RS0 are considered together because their value indicates the number of the active set (bank) of registers R0, R1...R7 engaged for data transfer. It means that the same register name is associated to other internal memory location. Physical memory addresses indicated by R0...R7 names depend on current configuration of bits RS1 and RS0 those are given in Table 1.2.

By default, after a microcontroller reset or switch the power on, the RS1 and RS0 bits are clear to zero, so the name of the R0 register allocated to memory cell with address 0, and R1 to the cell with address 1, etc. The bank

Table 1.2 Memory Space Allocated for R7...R0 Registers According to RS1 and RS0 Bits

RS1	RS0	Bank number	Memory addresses of internal RAM (as decimals)
0	0	0	0...7
0	1	1	8...15
1	0	2	16...23
1	1	3	24...31

switching mechanism is very useful as it shortens the program code. This will be illustrated by the following example.

Example 1.3: Save the contents of registers R0...R3 in the internal memory in order to use them for another task, and then restore their original value after finishing the task. The task can be completed in two ways. The first way – using the MOV instruction:

;Let's assume that the registers R0...R3 of bank 0 contain valid data

MOV 20h,R0   ;copy the value from R0 bank 0 to the internal memory cell at address 20h

MOV 21h,R1

MOV 22h,R2

MOV 23h,R3

;it is empty space left for the code that uses the R0...R3 bank 0 registers for another task

MOV R0,20h   ;copy the value from internal memory cell 20h to R0 bank 0

MOV R1,21h

MOV R2,22h

MOV R3,23h

The second way – using the bank switching mechanism:

;Let's assume that the registers R0...R3 of bank 0 contain valid data

SETB RS0   ;switch to bank no. 1

;it is empty space left for the code that uses the R0...R3 bank 0 registers for another task

CLR RS0   ;switch to bank no. 0

The meaning of the MOV, SETB, and CLR instructions used in Example 1.3 is explained in the next chapter. Do not to worry too much. It is not necessary to understand them at this stage. However, it is important to note that using the second solution results in shorter code. There are also other applications of the bank switching technique. One example can be implementation of function calling with parameter passing and local variables. The reader can be able to find more information searching the tutorials on learning assembly language.

 DID YOU KNOW?

In many processors, e.g. Microchip AVR, there are additional flags related to arithmetic or logical operations:

- zero (Z),
- negative (N),
- sign (S).

Their meaning is as follows:

- Z – set when result of an arithmetic or logical operation is zero,
- N – duplicated highest bit of the result, N = 1 indicates that the number is negative in the 2's complement format,
- S – set for negative result of an arithmetic operation, S = N $\oplus$ OV, which allows the correct interpretation of the condition by a jump instruction even if overflow occurs and result is incorrect.

Register – stores data or address.
Basic registers available for user are:

- accumulator A or ACC,
- register B,
- general purpose registers R0...R7 (x4 banks),
- 16-bits data pointer DPTR,
- program counter (instruction pointer) PC,
- stack pointer SP,
- flag register F / Program status word PSW.

Accumulator – read and write register used primarily for logical and arithmetic operations addressed by the most of instructions.

B – read and write universal register.

R0-R7 – read and write general purpose registers.

DPTR – read and write 16-bit register for addressing program or data memory, also available as pair of register DPH (high) and DPL (low).

Program counter/instruction pointer – read only register; addresses program memory, holds next instruction address or argument of current instruction, e.g. MOVC A,@A+PC.

Stack pointer – 8-bit register for reading and writing; it addresses the internal RAM with stack organization (LIFO – last in first out), it indicates the top of the stack, i.e. the address of the last written data.

The stack is used mainly for:

- temporary storing and restoring data with the PUSH and POP instructions to make register free to use by another instruction,
- store the value of the program counter PC during CALL instruction or handling an interrupt, i.e. jump/branch to the subroutine. The address from stack is stored back to the PC register with RET instruction or RETI for interrupt appropriately. It is resulting in returning to the next instruction of code after that causing a jump.

 DID YOU KNOW?

One of untypical applications of the stack [Null 2018, Gryś 2020] are:

- conversion of logical or arithmetic expressions from the classic infix to postfix form also known as the Reverse Polish Notation; benefit is that the parentheses are not necessary longer to force the priority of operations and less resources consumed,
- processing of logical or arithmetic expressions in postfix form,
- stack oriented data processed in Forth, Postscript language and some high-level language parsers,
- stack oriented registers of FPU unit of Intel processors.

1.3 UNDERSTANDING THE INSTRUCTION SET

Table 1.3 presents a list of 8051 microcontroller instructions, those using the following symbols:

Rn – R0...R7 registers of the currently selected register bank,

Ri – an internal data RAM location <0,255> addressed indirectly through R0 or R1,

address – 8-bit address of internal RAM memory,

Table 1.3 The 8051-Microprocessor Instruction Set Summary

Mnemonic	Arguments	Description	C	OV	AC
		Arithmetic operations			
ADD	A,Rn	A←A+Rn	X	X	X
	A,address	A←A+ [address]	X	X	X
	A,@Ri	A←A+[Ri]	X	X	X
	A,#data	A←A+data	X	X	X
ADDC	A,Rn	A←A+Rn+C	X	X	X
	A,address	A←A+ [address]+C	X	X	X
	A,@Ri	A←A+ [Ri]+C	X	X	X
	A,#data	A←A+data+C	X	X	X
SUBB	A,Rn	A←A−Rn−C	X	X	X
	A,address	A←A−[address]−C	X	X	X
	A,@Ri	A←A−[Ri]−C	X	X	X
	A,#data	A←A−data−C	X	X	X
INC	A	A←A+1			
	Rn	Rn←Rn+1			
	address	[address]←[address]+1			
	@Ri	[Ri]←[Ri]+1			
	DPTR	DPTR←DPTR+1			
DEC	A	A←A−1			
	Rn	Rn←Rn−1			
	address	[address]←[address]−1			
	@Ri	[Ri]←[Ri]−1			
MUL	AB	$B_{15...8}A_{7...0}$←A*B	0	If A*B>255, then OV←1	
DIV	AB	$A_{result}\ B_{remainder}$←A/B	0	If B = 0 before division then OV = 1	
DA	A	Decimal adjust accumulator after addition data in P-BCD: If $A_{3...0}$>9 or AC=1 then A←A+6 after that if $A_{7...4}$>9 or C=1 then A←A+60h	X		
		Logic operation			
ANL	A,Rn	A←A∩Rn			
	A,address	A←A∩[address]			
	A,@Ri	A←A∩[Ri]			
	A,#data	A←A∩data			
	address,A	[address]←[address]∩A			

(Continued)

Table 1.3 (Continued) The 8051-Microprocessor Instruction Set Summary

Mnemonic	Arguments	Description	C	OV	AC
	address,#data	[address]←[address]∩data			
ORL	A,Rn	A←A∪Rn			
	A,address	A←A∪[address]			
	A,@Ri	A←A∪[Ri]			
	A,#data	A←A∪data			
	address,A	[address]←[address]∪A			
	address,#-data	[address]←[address]∪A			
XRL	A,Rn	A←A⊕Rn			
	A,address	A←A⊕[address]			
	A,@Ri	A←A⊕[Ri]			
	A,#data	A←A⊕data			
	address,A	[address]←[address]⊕A			
	address,#-data	[address]←data⊕A			
CLR	A	A←0			
CPL	A	A←/A			
RL	A	Rotate accumulator left			
RLC	A	Rotate accumulator left through carry bit	X		
RR	A	Rotate accumulator right			
RRC	A	Rotate accumulator right through carry bit	X		
SWAP	A	$A_{7\ldots4} \leftrightarrow A_{3\ldots0}$			
Data transfer – internal memory					
MOV	A,Rn	A←Rn			
	A,address	A←[address]			
	A,@Ri	A←[Ri]			
	A,#data	A←data			
	Rn,A	Rn←A			
	Rn,address	Rn←[address]			
	Rn,#data	Rn←data			
	address,A	[address]←A			
	address,Rn	[address]←Rn			
	adress1,address2	[address1]←[address2]			

(Continued)

Table 1.3 (Continued) The 8051-Microprocessor Instruction Set Summary

Mnemonic	Arguments	Description	Flag		
			C	OV	AC
	address,@Ri	[address]←[Ri]			
	address,#data	[address]←data			
	@Ri,A	[Ri]←A			
	@Ri,address	[Ri]←[address]			
	@Ri,#data	[Ri]←data			
	DPTR,#data16	DPTR←data16			
XCH	A,Rn	A↔Rn			
	A,address	A↔ [address]			
	A,@Ri	A↔ [Ri]			
XCHD	A,@Ri	$A_{3...0}$↔ $[Ri]_{3...0}$			
PUSH	Address	[SP]←address, SP←SP+1			
POP	Address	[address]←[SP], SP←SP−1			

Data transfer – external memory and input/output devices

Mnemonic	Arguments	Description	C	OV	AC
MOVX	A,@Ri	A←[Ri]			
	A,@DPTR	A←[DPTR]			
	@Ri,A	[Ri]←A			
	@DPTR,A	[DPTR]←A			

Data transfer – program memory

Mnemonic	Arguments	Description	C	OV	AC
MOVC	A,@A+PC	A←[A+PC]			
	A,@A+DPTR	A←[A+DPTR]			

Single bit operation

Mnemonic	Arguments	Description	C	OV	AC
CLR	C	C←0	0		
	Bit	bit←0			
SETB	C	C←1	1		
	Bit	bit←1			
CPL	C	C←/C	X		
	Bit	bit←/bit			
ANL	C,bit	C←C∩bit	X		
	C,/bit	C←C∩/bit	X		
ORL	C,bit	C←C∪bit	X		
	C,/bit	C←C∪/bit	X		
MOV	C,bit	C←bit	X		
	bit,C	bit←C			

(Continued)

Table 1.3 (Continued) The 8051-Microprocessor Instruction Set Summary

Mnemonic	Arguments	Description	Flag		
			C	OV	AC
Unconditional jumps					
ACALL	address11	Subroutine call: $SP \leftarrow SP+1$, $[SP] \leftarrow PC_{7...0}$ $SP \leftarrow SP+1$, $[SP] \leftarrow PC_{15...8}$ $PC \leftarrow address11$			
LCALL	address16	Subroutine call $SP \leftarrow SP+1$, $[SP] \leftarrow PC_{7...0}$ $SP \leftarrow SP+1$, $[SP] \leftarrow PC_{15...8}$ $PC \leftarrow address16$			
RET		Return from subroutine: $PC_{15...8} \leftarrow [SP]$, $SP \leftarrow SP-1$ $PC_{7...0} \leftarrow [SP]$, $SP \leftarrow SP-1$			
RETI		Return from interrupt: $PC_{15...8} \leftarrow [SP]$, $SP \leftarrow SP-1$ $PC_{7...0} \leftarrow [SP]$, $SP \leftarrow SP-1$ Interrupts enabled with equal or less priority			
SJMP	Offset	$PC \leftarrow PC+offset$			
AJMP	address11	$PC_{10...0} \leftarrow address11$			
LJMP	address16	$PC \leftarrow address16$			
JMP	@A+DPTR	$PC \leftarrow A+DPTR$			
Conditional jumps					
JC	Offset	If $C=1$ then $PC \leftarrow PC+offset$			
JNC	Offset	If $C=0$ then $PC \leftarrow PC+offset$			
JB	Offset	If bit=1 then $PC \leftarrow PC+offset$			
JNB	Offset	If bit=0 then $PC \leftarrow PC+offset$			
JBC	bit,offset	If bit=1 then $PC \leftarrow PC+offset$ and bit$\leftarrow 0$			
JZ	Offset	If $A=0$ then $PC \leftarrow PC+offset$			
JNZ	Offset	If $A \neq 0$ then $PC \leftarrow PC+offset$			
CJNE	A,address,offset A,#data,offset Rn,#data,offset @ Ri,#data,offset	if $A \neq [address]$ then $PC \leftarrow PC+offset$ if $A \neq data$ then $PC \leftarrow PC+offset$ if $Rn \neq data$ then $PC \leftarrow PC+offset$ if $[Ri] \neq data$ then $PC \leftarrow PC+offset$	if $A \neq [address]$ then $C \leftarrow 1$ if $A \neq data$ then $C \leftarrow 1$ if $Rn \neq data$ then $C \leftarrow 1$ if $[Rn] \neq data$ then $C \leftarrow 1$		

(Continued)

Table 1.3 (Continued) The 8051-Microprocessor Instruction Set Summary

Mnemonic	Arguments	Description	Flag		
			C	OV	AC
DJNZ	Rn,offset address-s,offset	Rn←Rn−1 and if Rn≠0 then PC←PC+offset [address]←[address]−1 and if [address]≠0 then PC←PC+offset			
NOP		No operation			

address11 – 11-bit address of program memory. This argument is used by ACALL and AJMP instructions. The target of the CALL or JMP must lie within the same 2 KB range of addresses <−1024,1023>.

address16 – 16-bit address of program memory. This argument is used by LCALL and LJMP instructions.

data – 8-bits data,

data16 – 16-bits data,

bit – a direct addressed bit in internal data RAM or SFR memory (can be represented by name),

offset – a signed (two's complement) 8-bit offset <−128,127>,

X – value 0 or 1 as result of operation,

@... – memory addressed indirectly,

#... – a constant included in the instruction encoding,

/X – logical inversion of X,

[...] – the contents of the memory with address ... ,

KB – kilobyte, 2^{10} bytes = 1024 bytes,

BIN – binary code,

P-BCD – packed binary coded decimal.

1.4 ASSEMBLY LANGUAGE AND TOOLS

In previous subsection 1.3, the 8051-instruction set architecture was discussed in details. The program running on CPU is just a combination of instructions, addresses and operands translated to machine code understandable by microprocessor. To prepare and run program, we need some tools like: assembler, linker (optionally) and loader. The assembly language is a low-level programming language but assembler is a software responsible just for translating source code directly to the machine code or sometimes indirectly to the object code aimed to reuse in another program. A linker is a program merging one or more files generated earlier by a compiler or an assembler and combines them into a single executable file, library file or another 'object' file. Typically, one or more commonly used libraries are usually linked in by default. The linker also takes care of arranging the objects in a program's address space. This may involve relocating code that assumes a specific base address into another base. Relocating machine code may involve retargeting of absolute jump, MOV, load and store instructions. The loader, as its name suggests, is responsible for writing the final output file to the program memory of microprocessor. Sometimes it cooperates with bootloader located inside microprocessor memory and being a key feature of operating system kernel.

For assumed objectives of this book, the using of linker was not necessary. From this reason, we will concentrate only on assembling phase of the program creation needed to implement topics related to the title of a book. The assembly code was just written in plain text editor, translated to the machine code with final memory addresses and saved to the Intel HEX file format. To perform it, an attached assembler tool DSM51ASS.EXE was used but any freely available assembler dedicated to 8051 processor is also appropriate. All presented in this book listings were validated by testing output HEX output files with a real board equipped with 8051 CPU. Author is encouraging the potential readers to work (and maybe improve) with proposed algorithms and those implementation in code.

Basic features of the DSM51ASS assembler are as follows:

- assembles only a single input file (no linking phase),
- allows the use of complex arithmetic expressions similar to C language,
- allows the use of macros,
- allows the use the directives,
- checks the range of arguments.

A typical program line for this assembler looks like this:

[<label>] [<instruction>] [<operand>] [;<comment>]

The meaning of the individual fields of a program line is as follows:

    <label> – a symbol placed at the beginning of the line (the first character of the label must be the first character on the line). The label must start with a letter or the underscore '_', and may contain any combination of letters, numbers and underscores. If a label ends with a colon it is given a value that defines its

position in the source code (the address of an instruction from this program line). Labels (symbols) used with directives giving them a value are not terminated by a colon.

<instruction> – mnemonic of instruction, assembler directive or macro.

<operand> – the information required by the mnemonic, assembler directive or macro. Individual operands are separated by commas.

<comment> – all characters following a semicolon are treated as comments and ignored by the assembler during translation to machine code.

The instruction must be preceded by at least one whitespace character, e.g. space or tab character. There may be empty lines or lines containing only comments.

Arithmetic expressions are used to determine the value of parameters that require a numeric value. They consist of numbers and symbols (labels, names of constants or variables) combined with arithmetic operators. The syntax of DSM51ASS arithmetic expressions is very similar to that of the C language. It introduces some additional operators commonly used in assemblers, and changes the priority of bit operations – they are performed before comparison operations. The latter give the value 1 if the condition is true and 0 if it is false. Boolean operators (!,&&,||) treat any value other than zero as true and 0 as false. As a result of logical operations, we also get the value 1 or 0. In the DSM51ASS assembler, all calculations are performed on 32-bit signed numbers. This means that the value of an expression is calculated correctly as long as the intermediate results are within the range –2 147 483 648 to 2 147 483 647. Exceeding this range during the calculation is not signaled.

Symbols are represented by a string beginning with a letter or underscore '_' and consisting of any sequence of letters, numbers and underscores. The assembler recognizes the first 32 characters of a symbol.

Bit selection operator in bit-addressable registers is just a comma char:.n – the address of the specified bit in this register where 'n' is a digit from the range 0...7. If the given address is not the correct address of the bit-addressable register, an error is signaled.

For example:
START: CLR P1.7; comment

The numeric constant must start with a digit and ends with postfix for hexadecimal, octal and binary types. The H postfix used for hexadecimal numbers is equivalent to '0x' prefix in C-like commonly used notation and HEX subscript used in this book. The char is embraced by apostrophes. Some examples are presented below:

Type	Example
Decimal	123
Hexadecimal	0F28BH
Octal	76540
Binary	01010001B
Char	'A'

Operators that modify the value of the operand following it according to the priority of their execution:

() – parentheses determine the order in which actions are performed. There is no limit to the number of enclosing brackets used.

! – Boolean negation. Changes a value different from 0 to 0, and a value equal to 0 to 1.

~ – Bitwise negation. Changes all 32 bits in the operand to the opposite.

- – changes the sign of the operand to the opposite.

< – lowest byte of operand ('< operand' is equivalent to '(operand & FF_{HEX})' or '(operand % 256)').

> – higher 3 bytes of operand (an '> operand' is equivalent to '(operand >> 8)' or '(operand / 256)').

For example:

MOV A,#<((250-3)*2)   ;A=494-256=238=EE_{HEX}

Bitwise shift operators:

<< – Left shift. The operand to the left of this operator is shifted left by the number of bits specified by the operand to the right. The released bits are replaced by zeros.

For example:

MOV A,#(31+1)<<2;A=32*2*2=128238=80_{HEX}

The DSM51ASS assembler accepts the following directives:

DB – insert numeric and text values in the code, e.g. DB 'a',23,34H

DW – insert double-byte numeric values into the code, e.g. DW 2AE4H

EQU – define a constant, e.g. five EQU 00000101B

BIT – define constant of the bit type, e.g. my_bit BIT ACC.4

SET – define a variable, e.g. x_var SET 20H

IF/ELSE/ENDIF – start of conditional/alternative conditional/end of assembly block

ORG – set address for next block of code, e.g. ORG 10H

MACRO/ENDM – start/end of macro definition (sets of commands called by a single name), e.g.

MACRO

instruction1

instruction2

... .

instruction n

ENDM

They allow you to insert data into the program body, assign values to symbols, control the assembly flow and build macros. More details are provided below:

SET – to define a variable

Syntax: <symbol> SET <expression>

The symbol <symbol> is assigned the value of an expression. The symbol type is determined by the expression. The values defined by the SET directive can be modified any number of times by reusing the SET directive. Changing the symbol type during a subsequent assignment causes a warning to be generated.

IF/ELSE/ENDIF

Syntax – IF <expression> {code} ELSE {code_alter} ENDIF

ORG – set the address for the next block of code

Syntax – ORG <expression>

Sets the address for the code block following this directive. The address for the next processor instruction is determined by calculating the expression value. It is only possible to increment the current code address. Any attempt to decrease the address is signaled as an error. By default, the program code is starting from address zero.

MACRO/ENDM

Syntax – <name> MACRO <parameters>

A macro is a set of assembler instructions. The sequence of instructions following a line containing a MACRO directive, up to the nearest ENDM directive, forms a macro named <Name>. Once defined, this entire set can be included in the source code of a program by calling a macro, i.e. replacing its name. Calls to other macros may occur in the body of a macro, but the definition of another macro may not.

After above short information about assembler functionality and requirements, let's go to the next phase – creating the output files. The assembler generates the output file on the basis of source file, here with *.asm extension. It is then imported by a programmer to 'burn' the machine code into non-volatile memory, or is transferred to the operational RAM memory for loading and execution. All data required to run an application is included in binary file, typically with *.bin extension. This file contains raw code written to specific addresses, i.e. it represents program memory map. Unfortunately, it does not contain any mechanisms protecting file integrity. The loader responsible for writing the machine code to the program memory of microprocessor will not recognize the damage to the file or attempts to modify it. Moreover, if the code is distributed in different memory areas, it has to fill the empty addresses usually with the value 0 or 255 (0x00 or 0xFF) or left unattended. As a result, the volume of file is often oversized. In this book, we do not work with this kind of file but *.hex and *.lst files instead. The listing file (*.lst) is post-translating archive being a combination of source *.asm file and output *hex file. If errors or warnings have occurred they are also included and pointed to ease bug removing.

*.asm source file	*lst listing file
;******************	1 ;******************
;* An example of code *	2 ;* An example of code *
;******************	3 ;******************
n EQU 39H	4 0039 n EQU 39H
MOV A,#n	5 0000: 74 39 MOV A,#n
NOP	6 0002: 00 NOP
SJMP END	7 0003: 80 00 SJMP END
MUL AB	***** _ERROR 26: UNDEFINED SYMBOL *****
ADD A,C	8 0005: A4 MUL AB
	9 0006: 25 00 ADD A,C
	***** _ERROR 26: UNDEFINED SYMBOL *****

The same code with fixed bugs looks like this:

*.asm source file	*lst listing file	*.bin binary file	*.hex output file
;**************.*	1 ;************	00000000:	:0800000074390080-
An example of	2 ; * An example	74390003A425F0	03A425F00F
code	...		:00000001FF
*.*************	3 ;************		
n EQU 39H	4 0039 n EQU 39H		
MOV A,#n	5 0000: 74 39 MOV A,#n		
NOP	6 0002: 00 NOP		
SJMP END	7 0003: 80 03 SJMP END		
MUL AB	8 0005: A4 MUL AB		
ADD A,B	9 0006: 25 F0 ADD A,B		
END:	10 0008: END:		

Now, the *.bin or *.hex files were created. The structure of an Intel HEX file is very simple and will be described here as an example due to the fact that is commonly used during work not only with a family of 8051 processors. Firstly, it is a text file, because apart from digits it contains many colon characters (at least one if no code included). Each character, e.g. hexadecimal number '02', is encoded as ASCII two chars '0x30 0x32'. The way how to express the number values in various ways will be explained in Chapter 2. For example, the line can look like Figure 1.2

The line starts always with a char ':' as a *Start of a record*. The next field is single byte *Record length* giving the number of data bytes included in this line, maximum 256, so most often we see 0x10. Next one is the 16-bit *Address* field – starting address of program memory, where first byte is stored, here: 0x00. The address is always expressed as big endian value 0x0010 so for little endian convention, as used in this book, we have 0x0100 instead. The next field is Record type. If the data is just a code, then we have 0x00. Other values indicate the special meaning of the data, e.g. 0x01 – end of file. The last line of the file is special and always looks like this :00 0000 01 FF. The meaning and interpretation of *Data* field depends on the application. Mostly, it is just a machine code and some structure of data with strings, passwords to work with external devices like transmission terminal, LCD, touchscreen, calibrating factors of applied algorithms (e.g. digital instrument), look-up table (e.g. nonlinear functions, BIN to ASCII converter), etc. The line ends with a one byte of *Checksum* field. Its role is protection against loss of data integrity caused by errors during data transmission or storage and modification of file content. The way how to compute and check a checksum is quite easy, i.e. *Checksum*

:	10	0010	00	05 04 00 08 05 04 00 08 05 04 00 08 00 00 00 00	AD
Start of a record	Record length	Address	Record type	Data	Checksum

Figure 1.2 An example of correct line of hex file.

=256 –(sum modulo 256 of bytes in single line) as attached at the end of line during creating hex file. For line presented in Figure 1.2 and skipping bytes with value 0×00 we have 0×10+0×10+0×05+0×04+0×08+0×05+0×04 +0×08+0×05+0×04+0×08=0×53. Finally, 0×100-0×053=0×0AD=0×AD is a value of a checksum. After reading or receiving whole line, the checksum is evaluated again. Please note that the sum modulo 256 of all bytes together with the checksum should result in zero as confirmation of correct transmission. Unfortunately, the strength of data integrity protection is very low. Firstly, it is possible to get the same sum value for different combinations of number values or if, for example, errors cancelling each other out and occurred during a file transfer into processor memory, then *Checksum +error –error = Checksum.* It's worth adding that there are many HEX formats proposed by various companies such as Intel (HEX and HEX-32 file formats), Motorola (S-Record file format), and Tektronix (TEK HEX file format).

The ELF format is an executable file used to program the more powerful 32/64 processors like ARM family. It has a fixed segment structure. It contains headers, dedicated place for data, in our case: the program's machine code, additional place for data of text type. It is encrypted; therefore, it has the highest resistance to loss of integrity and is the recommended format for professional applications. Every modern hardware programming tool (sometimes called a bootloader) should handle with this format.

In next chapters of this book, we will present the *.lst listing files but also source assembly code is also available from the following link. https:// routledgetextbooks.com/textbooks/instructor_downloads/

Chapter 2

Numbers in Fixed-point Format

2.1 UNSIGNED NUMBERS

Performing arithmetic operations requires defining the understanding of the value of a number encoded in a bit word. Let's start with non-negative numbers at a beginning for ensuring clarity and simplicity of the information to be collected. Any A number is written using n digits in the integer part, and m digits in the fractional part according to the following format (Figure 2.1).

The point separates the integer part from the fractional part, i.e. the digits a_0 and a_{-1}. The term 'positional' means that the component a_i of number A is depending on its i-th position, where $i \in <n-1, -m>$. The value of A is taken as a weighted sum of digits. The ratio of the weights of two adjacent digits a_{i+1} and a_i, denoted here by the letter p, usually is a constant value, and is called the base or radix of the positional system. In practice, only systems with a positive base are used commonly to build computers, hence p = 2, 3, ..., ∞. Nevertheless, it is possible to imagine the number system with negative base or even variable, depending on the digit position. The term 'fixed' emphasizes the observation that the value of a number can be expressed in only one way. By limiting the considerations to non-negative numbers with a positive base, the value of the number is determined from the relationship (2.1):

$$A = a_{n-1} \cdot p^{n-1} + a_{n-2} \cdot p^{n-2} + \ldots + a_1 \cdot p + a_0 + a_{-1} \cdot p^{-1} + \ldots + a_{-m} \cdot p^{-m}$$

$$= \sum_{i=-m}^{n-1} a_i \cdot p^i \tag{2.1}$$

where a_i – i-th digit, p – base (radix).

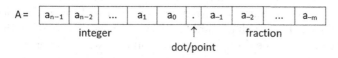

A =	a_{n-1}	a_{n-2}	...	a_1	a_0	.	a_{-1}	a_{-2}	...	a_{-m}
	integer					↑	fraction			
						dot/point				

Figure 2.1 A fixed-point format of number.

DOI: 10.1201/9781003363286-2

For the graphical representation of digits at a given base, the set of Arabic numerals and Roman letters are used, that amount depends on the assumed base. More information about the fixed-point format and its properties can be found in many books, e.g. [Pochopień 2012, Scott 1985]. There are, of course, other non-positional ways of coding numbers, e.g. the residual system mentioned in [Biernat 2007, Parhami 2010].

Example 2.1:

    a. $p = 2$ – binary system BIN, digits are in the range $\{0,1\}$, e.g.:

$$101.01 = 1 \cdot 2^2 + 0 \cdot 2^1 + 1 \cdot 2^0 + 0 \cdot 2^{-1} + 1 \cdot 2^{-2}$$

    b. $p = 10$ – decimal system DEC, digits are in the range $\{0,1,2, \ldots ,9\}$, e.g.:

$$194.23 = 1 \cdot 10^2 + 9 \cdot 10 + 4 + 2 \cdot 10^{-1} + 3 \cdot 10^{-2}$$

    c. $p = 16$ – hexadecimal (hexagonal) system HEX, digits are in the range $\{0,1,2, \ldots,9,A,B, \ldots, F\}$, e.g.:

$$A4.B = A \cdot 16^1 + 4 \cdot 16^0 + B \cdot 16^{-1}$$

With regard to $p = 2$, the term natural binary code or binary code is used. In order to maintain full formalism, the term 'natural' should also be applied to the other two systems, $p = 10$ and $p = 16$, which, however, is not commonly done involving misinterpretation sometimes if different 'binary systems' are mixed. Apart from the above-mentioned systems, infinitely many others can be defined, but systems with the base $p = 2, 10, 16$ as shown above have the greatest practical significance. The octal system $p = 8$ is also mentioned in many books and tutorials but in my opinion, its usefulness is rather insignificant (during many years of practice, the author did not have the opportunity to use it), so we will not devote any more attention to it. Of course, this issue may be matter for further discussion trying to highlight possible advantages of these systems with basis, e.g. $p = 3, 4, 7$, as shown in [Pankiewicz 1985].

The decimal system has reached widespread acceptance in everyday life – probably due to the anatomy of the human hand and number of fingers. The binary system corresponds to the two-state model of information processed or stored by computers. Two states relate to two distinguished values of electric voltage or current in technical realizations of electronic devices. Given the fact that number 16 is a natural power of 2, the hexadecimal format can be thought as much compact way of expressing binary numbers. There is a general rule: the greater the value of p, the more different numbers can be written with the same number of digits thus higher density. With n and m digits, we can express p^{n+m} combinations of different values. The smallest difference between the two values is called the resolution and is p^{-m}.

Example 2.2: Maximum numbers for n = 3, m = 1 and p = 2, 10, 16 are, respectively:

   a. p = 2, $111.1_{BIN} = 7.5_{DEC} = 2^3 - 2^{-1}{}_{DEC}$
   b. p = 10, $999.9_{DEC} = 10^3 - 10^{-1}{}_{DEC}$
   c. p = 16, $FFF.F_{HEX} = 16^3 - 16^{-1}{}_{DEC}$

Example 2.3: Minimum non-zero numbers for n = 2, m = 3 and p = 2, 10, 16, are, respectively:

   a. p = 2, $00.001_{BIN} = 2^{-3}{}_{DEC}$
   b. p = 10, $00.001_{DEC} = 10^{-3}{}_{DEC}$
   c. p = 16, $00.001_{HEX} = 16^{-3}{}_{DEC}$

Exercise 2.1: Determine the maximal numbers for n = 1, m = 2 and p = 2, 10, 16.

In practice, one can often encounter the problem of expressing a number using different base. The easy way of conversion is just derived from the property of Eq. (2.1). The digits are constant weights depending on position in digit field. Therefore, it is convenient to use an auxiliary template as presented in Example 2.4, whereby we just make conversion between the decimal and binary systems. The hexadecimal number is easy to rich from binary representation by applying 8421 weights for every 4 bits, individually. A general note about the accepted number expression convention in this book is as follows: if no base value is explicitly given, it refers to the decimal system!

Example 2.4: Converting DEC→BIN→HEX:

   a. 23_{DEC}→

128	64	32	16	8	4	2	1	
0	0	0	1	0	1	1	1	BIN

$\updownarrow$ $\updownarrow$

8	4	2	1	8	4	2	1	
1				7				HEX

b. 15.75_{DEC}→

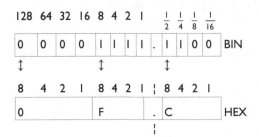

128	64	32	16	8	4	2	1		$\frac{1}{2}$	$\frac{1}{4}$	$\frac{1}{8}$	$\frac{1}{16}$	
0	0	0	0	1	1	1	1	.	1	1	0	0	BIN

8	4	2	1	8	4	2	1	8	4	2	1	
0				F			.	C				HEX

Exercise 2.2: Represent the assumed numbers in other two formats:

a. 246.5_{DEC} → $?...?_{BIN}$ → $?...?_{HEX}$
b. $3E.4_{HEX}$ → $?...?_{BIN}$ → $?...?_{DEC}$
c. 10110011.0010_{BIN} → $?...?_{HEX}$ → $?...?_{DEC}$

Applying template for fractional part, we see that it requires the addition of fractions with different denominator values. Contrary, it is recommended to work as explained in Example 2.5. The routine consists in repeating the multiplication of the fraction by 2 (if converting to binary system). When determining the successive bits of the binary fraction, the integer part (single bit) of the multiplication result is not considered.

Example 2.5: Converting the fraction DEC → fraction BIN → fraction HEX:

0.8125·2	0.75I·2
1.6250·2	1.502·2
1.25·2	1.004·2
0.5·2	0.008·2
1.0·2	0.016·2
0.0·2	0.032·2
	0.064·2
	0.128·2
	0.256·2
	0.512·2
	1.024·2
	0.048·2
	0.096·2
	...

0.8125_{DEC} → 0.1101_{BIN} → 0.751_{DEC} → ≈0.1100 0000 0100_{BIN} →
$0.D_{HEX}$ $\approx0.C04_{HEX}$

Checking:

0.5000	0.500
0.2500	0.250
+ 0.0625	+ ≈ 0.001
0.8125$_{DEC}$	0.751$_{DEC}$

The same is done for simple fractions as shown in the next example. The algorithm is universal and can be used also for a decimal system or any base.

Example 2.6: Converting rational number 5/6$_{DEC}$ to the fraction as DEC, BIN and HEX:

(0 + 5/6)·10	(0 + 5/6)·2
(8 + 2/6)·10	(1 + 4/6)·2
(3 + 2/6)·10	(1 + 2/6)·2
(3 + 2/6)·10	(0 + 4/6)·2
...	(1 + 2/6)·2
	(0 + 4/6)·2
	(1 + 2/6)·2
	(0 + 4/6)·2
	(1 + 2/6)·2
	...

5/6DEC → ≈ 5.833 ...$_{DEC}$ 5/6DEC →   0.110101(01)$_{BIN}$ →   ≈ 0.D5$_{HEX}$

Exercise 2.3: Convert numbers to the HEX and BIN fractions:

a. 0.63$_{DEC}$
b. 11/9$_{DEC}$
c. 3/5$_{DEC}$
d. 1/128$_{DEC}$

 INTERESTING FACTS!

The expansion of a rational number for the radix p is always finite or periodic, e.g.:

$$3/8_{DEC} = 0.375_{DEC} = 0.011_{BIN} - \text{finite}$$

$$2/3_{DEC} = 0.(6) = 0.(10)_{BIN} - \text{periodic}$$

Some external devices enabling human-machine interface, such as the alphanumeric LCD display, are processing data in other formats. In the context of arithmetic, the 8421 BCD decimal code and the ASCII codes are important. Most processors can work with these codes delivering the correct result thanks to proper instruction of correction. The examples are the processors compatible with the instruction list of the ancestor of modern CPU – the Intel 8086 processor. The discussed instructions are DAA, DAS, AAA, AAS, AAM and AAD [Irvine 2003]. Unfortunately, the 8051 micro-controller that we work in this book, only performs the decimal correction DAA after adding (see Table 1.3). For 8421 BCD format, each decimal digit is represented by four or eight binary digits, with such combinations of bits to express decimal values only, i.e. in the range 0–9. The 4-bit BCD is called packed BCD, denoted as P-BCD in the book, and 8-bit BCD format with four padding zeros is called unpacked BCD and denoted as UP-BCD.

Example 2.7: Number coded as P-BCD and UP-BCD and its decimal representation as DEC:

a. $00010010000000011.00001000_{UP-BCD} = 10010011.1000_{P-BCD} = 93.8_{DEC}$
b. $000000100000001000000110_{UP-BCD} = 001001000110_{P-BCD} = 246_{DEC}$

Exercise 2.4: Convert DEC numbers to P-BCD and UP-BCD:

a. 479.12_{DEC}
b. 0.03_{DEC}
c. 8.9_{DEC}
d. 123_{DEC}

Another well-known BCD code is the not-weighted BCD code Excess 3. It is formed by adding the number 3 to the 8421 BCD code. This apparent inconvenience carries the beneficial property of self-completion, which is important for the simplicity of hardware implementation of arithmetic circuits. However, it is not commonly used by ALU, so we will not devote more attention to it. The ASCII code assigns sequential numbers to graphic characters and control symbols of peripheral devices, most of which are obsolete today. In some languages, e.g. C/C++, only a few of them are meaningful, e.g. 0 – null for terminating a text string, 10 – LF (line feed) and 13 – CR (carrier return) in *printf()* function for formatting the output. The ASCII code array contains a rich set of graphical characters, including the Latin and Greek alphabets, punctuation marks, mathematical symbols and semi graphics [Irvine 2003]. It is mainly used to store the appearance of characters in the persistent memory of LCD alphanumeric display controllers used in various control, measurement and monitoring devices, i.e. oven controllers, beverage vending machines, home appliances and many others. With the development of the Internet and the need to represent alphabets of many languages, ASCII code was absorbed by Unicode (UTF-8) and Universal Character Set (UCS, ISO 10646) and is now represented in the same positions in the bigger table, occupying codes 0–127. Operating on the full ASCII code table is not necessary in the context of arithmetic, since we are only interested in the way how the decimal digits are encoded. In fact, it is very simple and it is enough to prefix the encoded decimal digit 0...9 with 3_{DEC} coded binary as 0011. The digit code is then given as a hexadecimal number. It can also be expressed in the zero-one system. The comma sign is encoded as $2E_{HEX}$, or 00101110_{BIN}. For obvious reasons, these values should not be interpreted do verbally as hexadecimal or binary numbers according to Eq. (2.1).

Example 2.8: ASCII codes of decimal numbers:

   a. $36_{ASCII} = 6_{DEC}$
   b. $00110010\ 00101110\ 00110101_{ASCII} = 2.5_{DEC}$

Exercise 2.5: Convert DEC number to ASCII code:

   a. 361.82_{DEC}
   b. 36.18_{DEC}
   c. 0.45_{DEC}
   d. 97.1_{DEC}

2.2 CONVERSION OF UNSIGNED NUMBER TO ANOTHER FORMAT

This chapter will show you how to convert numbers programmatically. The subroutines have been written in the assembly code of the 8051 family microcontroller. The code, the meaning of instructions, the content of registers A, B and sometimes R0 have been presented in tables. In order to facilitate the analysis of the code, its operation is illustrated on real numerical values. Shaded values of A, B and R0 presented in tables indicate the input and output values of the algorithm. Subsequent rows of the table show the current contents of the registers after running line of code. The meaning of the numeric argument suffixes is as follows: h or H – hexadecimal notation, B – binary, none – decimal. Additionally, a full listing of algorithm is provided. It consists of two main parts: instructions responsible for writing the input arguments to the registers of the microcontroller and subroutine starting with a label name and ending with a RET instruction. Each line of the listing starts with line number and then optionally there can be such elements as hexadecimal value of the constant and its symbolic name, address of the program memory cell where the first byte of the machine code is placed (the address is followed by colon), label ended by colon again, instruction mnemonic with operands and comment preceded by semicolon sign. Each listing ends with a line with the comment like this '--- end of file ---'.

2.2.1 Conversion BIN to P-BCD for A < 100$_{DEC}$

- input number in A,
- output number in A,
- exemplary value: $54_{DEC} = 00110110_{BIN} = 01010100_{P\text{-}BCD}$.

The algorithm uses the properties of the arithmetic instruction DIV. Dividing the input argument by the number 10_{DEC}, you get tens in register A and unities in register B. The next SWAP and ORL instructions convert the obtained result, interpreted as a number in UP-BCD, into a packed BCD form. The algorithm returns a correct result for an input number from the range 0_{DEC} and 255_{DEC} (Table 2.1).

Table 2.1 Implementation in Code and Intermediate Results

Code	Description	A	B
MOV A,#54	Let A be 54$_{DEC}$	00110110	
MOV B,#0Ah	Let B be 10$_{DEC}$ given as hexadecimal	00110110	00001010
DIV AB	Divide A by B, A – result, B – reminder	00000101	00000100
SWAP A	Change nibbles A$_{7\ldots4}$ ↔ A$_{3\ldots0}$	01010000	00000100
ORL A,B	Perform A∪B and save result	01010100	00000100

1	;***		
2	;*	Conversion BIN to P-BCD for n<100DEC*	
3	;***		
4	0036	n EQU 54	;n=54 DEC
5			
6	0000: 74 36	MOV A,#n	;input a number
7	0002: 12 00 07	LCALL BIN_P_BCD100	
8			;result in A
9	0005: 80 FE	STOP: SJMP STOP	;infinite loop
10	;--		
11	0007:	BIN_P_BCD100:	
12	0007: 75 F0 0A	MOV B,#10	;let B be 10DEC
13	000A: 84	DIV AB	;divide A by B, A-tens, B-unities
14	000B: C4	SWAP A	;change nibbles A7..4<->A3..0
15	000C: 45 F0	ORL A,B	;perform (A OR B)
16	000E: 22	RET	
17	;— end of file —		

2.2.2 Conversion BIN to P-BCD for A < 256$_{DEC}$

- input: number in A,
- output: number in A – hundreds, $B_{7...4}$ – tens, $B_{3...0}$ – unities,
- exemplary value: $153_{BIN} = 10011001_{BIN} = 000101010011_{P\text{-}BCD}$.

By separating out the hundreds the algorithm can be used for numbers between 0_{DEC} and 255_{DEC} (Table 2.2).

Table 2.2 Implementation in Code and Intermediate Results

Code	Description	A	B	R0
MOV A,#153	Input a number	10011001		
MOV B,#100	Let B be 100$_{DEC}$	10011001	01100100	
DIV AB	Divide A by B, A – hundreds, B – rest	00000001	00110101	
XCH A,B	Exchange A ↔ B	00110101	00000001	
MOV R0,B	Save hundreds to R0	00110101	00000001	00000001
MOV B,#0Ah	Let B be 10$_{DEC}$	00110101	00001010	00000001
DIV AB	Divide A by B, A – tens, B – unities	00000101	00000011	00000001
SWAP A	Change nibbles $A_{7...4}$ ↔ $A_{3...0}$	01010000	00000011	00000001
ORL A,B	Perform A∪B and save result	01010011	00000011	00000001
MOV B,R0	Load 'hundreds' to B	01010011	00000001	00000001
XCH A,B	Exchange A ↔ B	00000001	01010011	00000001

```
 1      ;******************************************************************
 2      ;*                    Conversion BIN to P-BCD for n<256DEC*
 3      ;******************************************************************
 4      0099                    n EQU 153        ;n=153 DEC
 5
 6      0000: 74 99             MOV A,#n         ;input a number
 7      0002: 12 00 07          LCALL
                                BIN_P_BCD256
 8                                               ;result in A-hundreds,
                                                  B-tens and unities
 9      0005: 80 FE             STOP: SJMP STOP
10      ;---------------------------------------------------------------------
11      0007:                   BIN_P_BCD256:
12      0007: 75 F0 64          MOV B,#100       ;let B be 100DEC
13      000A: 84                DIV AB           ;divide A by B, A-hundreds, B-
                                                  rest
14      000B: C5 F0             XCH A,B          ;exchange A<->B
15      000D: A8 F0             MOV R0,B         ;save hundreds to R0
```

16	000F: 75 F0 0A	MOV B,#10	;let B be 10DEC
17	0012: 84	DIV AB	;divide A by B, A-tens, B-unities
18	0013: C4	SWAP A	;change nibbles A7..4<->A3..0
19	0014: 45 F0	ORL A,B	;perform (A OR B) and save result
20	0016: 88 F0	MOV B,R0	;load hundreds to B
21	0018: C5 F0	XCH A,B	;exchange A<->B
22	001A: 22	RET	
23	;--- end of file ---		

2.2.3 Conversion BIN to UP-BCD for A < 100$_{DEC}$

- input: number in A,
- output: number in A – tens, B – unities,
- exemplary number: $67_{DEC} = 01000011_{BIN} = 00000110\ 00000111_{UP\text{-}BCD}$.

The algorithm returns a correct result for an input number between 0_{DEC} and 255_{DEC} (Table 2.3).

Table 2.3 Implementation in Code and Intermediate Results

Code	Description	A	B
MOV A,#67	Input a number	01000011	
MOV B,#10	Let B be 10$_{DEC}$	01000011	00001010
DIV AB	Divide A by B, A – tens, B – unities	00000110	00000111

1	;***
2	;* Conversion BIN to UP-BCD for n<100DEC*
3	;***
4	0036	n EQU 54	;n=54 DEC
5	
6	0000: 74 36	MOV A,#n	;input a number
7	0002: 12 00 07	LCALL BIN_UP_BCD100	
8			;result in A-tens, B-unities
9	0005: 80 FE	STOP:	SJMP STOP
10	;---
11	0007:	BIN_UP_BCD100:	
12	0007: 75 F0 0A	MOV B,#10	;let B be 10DEC
13	000A: 84	DIV AB	;divide A by B
14	000B: 22	RET	
15	;--- end of file ---		

2.2.4 Conversion BIN to UP-BCD for A < 256$_{DEC}$

- input: number in A,
- output: number in R0 – hundreds, A – tens, B – unities,
- exemplary value: $153_{DEC} = 10011001_{BIN} = 0000000100000101000$ 00011_{UP-BCD}.

The algorithm returns a correct result for an input number between 0_{DEC} and 255_{DEC} (Table 2.4).

Table 2.4 Implementation in Code and Intermediate Results

Code	Description	R0	A	B
MOV A,#153	Input a number		10011001	
MOV B,#100	Let B be 100$_{DEC}$		10011001	01100100
DIV AB	Divide A by B, A – hundreds, B – rest		00000001	00110101
XCH A,B	Exchange A ↔ B		00110101	00000001
MOV R0,B	Save hundreds to R0	00000001	00110101	00000001
MOV B,#0Ah	Let B be 10$_{DEC}$	00000001	00110101	00001010
DIV AB	Divide A by B, R0 – hundreds, A – tens, B – unities	00000001	00000101	00000011

```
 1      ;*******************************************************************
 2      ;*              Conversion BIN to UP-BCD for n<256DEC*
 3      ;*******************************************************************
 4      0099            n EQU 153                  ;n=153 DEC
 5
 6      0000: 74 99     MOV A,#n                   ;input a number
 7      0002: 12 00 07  LCALL NKD_UP_BCD256
 8                                                 ;result in R0-hundreds, A-
                                                    tens, B-unities
 9      0005: 80 FE     STOP:          SJMP STOP
10      ;--------------------------------------------------------------------
11      0007:           NKD_UP_BCD256:
12      0007: 75 F0 64  MOV B,#100                 ;let B be 100DEC
13      000A: 84        DIV AB                     ;divide A by B, A-
                                                    hundreds, B-rest
14      000B: C5 F0     XCH A,B                    ;exchange A<->B
15      000D: A8 F0     MOV R0,B                   ;save hundreds to R0
16      000F: 75 F0 0A  MOV B,#10                  ;let B be 10DEC
17      0012: 84        DIV AB                     ;divide A by B
18      0013: 22        RET
19      ;--- end of file ---
```

2.2.5 Conversion BIN to ASCII for A < 100$_{\text{DEC}}$

- input: number in A,
- output: number in A – tens, B – unities,
- exemplary value: $67_{\text{DEC}} = 01000011_{\text{BIN}} = 00110110\ 00110111_{\text{ASCII}}$.

The algorithm returns a correct result for an input number between 0_{DEC} and 255_{DEC} (Table 2.5).

Table 2.5 Implementation in Code and Intermediate Results

Code	Description	A	B
MOV A,#67	Input a number	01000011	
MOV B,#10	Let B be 10$_{\text{DEC}}$	01000011	00001010
DIV AB	Divide A by B, A – tens, B – unities	00000110	00000111
XCH A,B	Exchange A ↔ B	00000111	00000110
ADD A,#30h	Add 30$_{\text{HEX}}$ to A	00110111	00000110
XCH A,B	Exchange A ↔ B	00000110	00110111
ADD A,#30h	Add 30$_{\text{HEX}}$ to A	00110110	00110111

```
 I      ;***************************************************************
 2      ;*                  Conversion BIN to ASCII for n<100DEC*
 3      ;***************************************************************
 4      0043              n EQU 67                ;n=67 DEC
 5
 6      0000: 74 43       MOV A,#n                ;input a number
 7      0002: 12 00 07    LCALL BIN_ASCII100
 8                                                ;result in A-tens, B-unities
 9      0005: 80 FE       STOP:          SJMP STOP
10              ;------------------------------------------------------------------------
11      0007:             BIN_ASCII100:
12      0007: 75 F0 0A    MOV B,#10               ;let B be 10DEC
13      000A: 84          DIV AB                  ;divide A by B, A-tens,
                                                     B-unities
14      000B: C5 F0       XCH A,B                 ;exchange A<->B
15      000D: 24 30       ADD A,#30H              ;add 30h to A
16      000F: C5 F0       XCH A,B                 ;exchange A<->B
17      0011: 24 30       ADD A,#30H              ;add 30h to A
18      0013: 22          RET
19      ;--- end of file ---
```

2.2.6 Conversion BIN to ASCII for A < 256$_{DEC}$

- input: number in A,
- output: number in R0 – hundreds, A – tens, B – unities,
- exemplary value: $153_{BIN} = 10011001_{BIN} = 001100010011010100$
  110011_{ASCII}.

The algorithm returns a correct result for an input number between 0_{DEC} and 255_{DEC} (Table 2.6).

Table 2.6 Implementation in Code and Intermediate Results

Code	Description	R0	A	B
MOV A,#153	Input a number		10011001	
MOV B,#100	Let B be 100$_{DEC}$		10011001	01100100
DIV AB	Divide A by B, A – hundreds, B – rest		00000001	00110101
ADD A,#30h	Add 30$_{HEX}$ to A		00110001	00110101
XCH A,B	Exchange A ↔ B		00110101	00110001
MOV R0,B	Save hundreds to R0	00110001	00110101	00110001
MOV B,#0Ah	Let B be 10$_{DEC}$	00110001	00110101	00001010
DIV AB	Divide A by B, A – tens, B – unities	00110001	00000101	00000011
XCH A,B	Exchange A ↔ B	00110001	00000011	00000101
ADD A,#30h	Add 30$_{HEX}$ to A	00110001	00110011	00000101
XCH A,B	Exchange A ↔ B	00110001	00000101	00110011
ADD A,#30h	Add 30$_{HEX}$ to A	00110001	00110101	00110011

```
 I    ;*******************************************************************
 2    ;*              Conversion BIN to ASCII for n<256DEC*
 3    ;*******************************************************************
 4    0099            n EQU 153           ;n=153 DEC
 5
 6    0000: 74 99      MOV A,#n            ;input a number
 7    0002: 12 00 07   LCALL BIN_ASCII256
 8                                         ;result in R0-hundreds,
                                            A-tens, B-unities
 9    0005: 80 FE      STOP:      SJMP STOP
10               ;----------------------------------------------------------------
11    0007:           BIN_ASCII256:
12    0007: 75 F0 64   MOV B,#100          ;let B be 100DEC
13    000A: 84         DIV AB              ;divide A by B, A-hundreds,
                                            B-rest
```

14	000B: 24 30	ADD A,#30H	;add 30h to A
15	000D: C5 F0	XCH A,B	;exchange A<->B
16	000F: A8 F0	MOV R0,B	;save hundreds to R0
17	0011: 75 F0 0A	MOV B,#10	;let B be 10DEC
18	0014: 84	DIV AB	;divide A by B, A-tens, B-unities
19	0015: C5 F0	XCH A,B	;exchange A<->B
20	0017: 24 30	ADD A,#30h	;add 30h to A
21	0019: C5 F0	XCH A,B	;exchange A<->B
22	001B: 24 30	ADD A,#30h	;add 30h to A
23	001D: 22	RET	
24	;--- end of file ---		

2.2.7 Conversion P-BCD to BIN

- input: number in A,
- output: number in A,
- exemplary number: $01010100_{P\text{-}BCD} = 00110110_{BIN}$ (see Table 2.7).

Table 2.7 Implementation in Code and Intermediate Results

Code	Description	A	B	R0
MOV A,#54h	Input a number	01010100		
MOV R0,A	Make a copy to R0	01010100		01010100
ANL A,#0F0h	Clear lower nibble of $A_{3...0}$	01010000		01010100
SWAP A	Change nibbles $A_{7...4} \leftrightarrow A_{3...0}$	00000101		01010100
MOV B,#0Ah	Let B be 10_{DEC}	00000101	00001010	01010100
MUL AB	Multiply A by B	00110010	00000000	01010100
MOV B,R0	Load original number to B	00110010	01010100	01010100
ANL B,#0Fh	Clear higher nibble of $B_{7...4}$	00110010	00000100	01010100
ADD A,B	Add B to A	00110110	00000100	01010100

```
 1   ;***************************************************************
 2   ;*              Conversion P-BCD to BIN    *
 3   ;***************************************************************
 4   0054          n EQU 54h             ;n=54 P-BCD
 5
 6   0000: 74 54    MOV A,#n             ;input a number
 7   0002: 12 00 07 LCALL P_BCD_BIN
 8                                       ;result in A
 9   0005: 80 FE    STOP:    SJMP STOP
10   ;-------------------------------------------------------------
11   0007:          P_BCD_BIN:
```

```
12   0007: F8          MOV R0,A              ;make a copy to R0
13   0008: 54 F0        ANL A,#0F0H          ;clear lower nibble of A3..0
14   000A: C4           SWAP A               ;change nibbles A7..4<->A3..0
15   000B: 75 F0 0A      MOV B,#0AH           ;let B be 10DEC
16   000E: A4           MUL AB               ;multiply A by B
17   000F: 88 F0         MOV B,R0             ;load original number to B
18   0011: 53 F0 0F      ANL B,#0FH           ;clear higher nibble of B7..4
19   0014: 25 F0         ADD A,B              ;add B to A
20   0016: 22           RET
21   ;--- end of file ---
```

2.2.8 Conversion P-BCD to UP-BCD

- input: number in A,
- output: number in A – tens, B – unities,
- exemplary value: $01000111_{P\text{-}BCD} = 00000100\ 0000111_{UP\text{-}BCD}$ (see Table 2.8).

Table 2.8 Implementation in Code and Intermediate Results

Code	Description	A	B
MOV A,#47h	Input a number	01000111	
MOV B,A	Make a copy to B	01000111	01000111
ANL A,#0F0h	Clear lower nibble of $A_{3...0}$	01000000	01000111
ANL B,#0Fh	Clear higher nibble of $B_{7...4}$	01000000	00000111
SWAP A	Change nibble $A_{7...4} \leftrightarrow A_{3...0}$	00000100	00000111

```
1    ;*************************************************************
2    ;*              Conversion P-BCD -> UP-BCD*
3    ;*************************************************************
4    0047                n EQU 47h             ;n=47 P-BCD
5
6    0000: 74 47          MOV A,#n              ;input a number
7    0002: 12 00 07       LCALL P_BCD_UP_BCD
8
9    0005: 80 FE          STOP:                 SJMP STOP
10   ;--------------------------------------------------------------
11   0007:               P_BCD_UP_BCD:
12   0007: F5 F0          MOV B,A               ;make a copy to B
13   0009: 54 F0          ANL A,#0F0h           ;clear lower nibble of A3..0
14   000B: 53 F0 0F       ANL B,#0Fh            ;clear higher nibble of B7..4
15   000E: C4             SWAP A                ;change nibbles A7..4<->A3..0
```

16			;result in A-tens, B-unities
17	000F: 22	RET	
18	;--- end of file ---		

2.2.9 Conversion P-BCD to ASCII

- input: number in A,
- output: number in A – tens, B – unities,
- exemplary number: $01000101_{P\text{-}BCD} = 00110100\ 00110101_{ASCII}$ (see Table 2.9).

Table 2.9 Implementation in Code and Intermediate Results

Code	Description	A	B
MOV A,#45h	Input a number	01000101	
MOV B,A	Make a copy to B	01000101	01000101
ANL A,#0Fh	Clear higher nibble of $A_{7...4}$	00000101	01000101
ADD A,#30h	Add 30_{HEX} to A	00110101	01000101
XCH A,B	Exchange A $\leftrightarrow$ B	01000101	00110101
ANL A,#0F0h	Clear lower nibble of $A_{3...0}$	01000000	00110101
SWAP A	Change nibbles $A_{7...4} \leftrightarrow A_{3...0}$	00000100	00110101
ADD A,#30h	Add 30_{HEX} to A	00110100	00110101

```
 1        ;*******************************************************************
 2        ;* Conversion P-BCD to ASCII *
 3        ;*******************************************************************
 4        0045           n EQU 45h              ;n=45 P-BCD
 5
 6        0000: 74 45     MOV A,#n              ;input a number
 7        0002: 12 00 07  LCALL P_BCD_ASCII
 8                                              ;result in A-tens, B-unities
 9        0005: 80 FE     STOP:          SJMP STOP
10        ;------------------------------------------------------------------
11        0007:           P_BCD_ASCII:
12        0007: F5 F0     MOV B,A               ;make a copy to B
13        0009: 54 0F     ANL A,#0Fh            ;clear a higher nibble of A7..4
14        000B: 24 30     ADD A,#30h            ;add 30h to A
15        000D: C5 F0     XCH A,B               ;exchange A<->B
16        000F: 54 F0     ANL A,#0F0h           ;clear lower nibble of A3..0
17        0011: C4        SWAP A                ;change nibbles of A7..4<->A3..0
18        0012: 24 30     ADD A,#30h            ;add 30h to A
19        0014: 22        RET
20        ;--- end of file ---
```

2.2.10 Conversion UP-BCD to BIN

- input: number in A – tens, B – unities,
- output: number in A,
- exemplary value: $00001001\ 00000110_{UP\text{-}BCD} = 01100000_{BIN}$ (see Table 2.10).

Table 2.10 Implementation in Code and Intermediate Results

Code	Description	A	B	R0
MOV A,#09h	Input a first number (tens)	00001001		
MOV B,#06h	Input a second number (unities)	00001001	00000110	
MOV R0,B	Make a copy to B	00001001	00000110	00000110
MOV B,#10	Let B be 10$_{DEC}$	00001001	00001010	00000110
MUL AB	Multiply A by B	01011010	00000000	00000110
ADD A,R0	Add R0 to A	01100000	00000000	00000110

```
 1   ;************************************************************************
 2   ;*                    Conversion UP-BCD to BIN*
 3   ;************************************************************************
 4   0009              n EQU 09h
 5   0006              m EQU 06h              ;{nm} 0906 UP-BCD = 96 DEC
 6   0000: 74 09       MOV A,#09h             ;input a first number
 7   0002: 75 F0 06    MOV B,#06h             ;input a second number
 8   0005: 12 00 0A    LCALL UP_BCD_BIN
 9                                            ;result in A
10   0008: 80 FE       STOP:       SJMP STOP
11        ;--------------------------------------------------------------------
12   000A:             UP_BCD_BIN:
13   000A: A8 F0       MOV R0,B               ;make a copy to R0
14   000C: 75 F0 0A    MOV B,#10              ;let B be 10DEC
15   000F: A4          MUL AB                 ;multiply A by B
16   0010: 28          ADD A,R0               ;add R0 to A
17   0011: 22          RET
18   ;--- end of file ---
```

2.2.11 Conversion UP-BCD to P-BCD

- input: number in A – tens, B – unities,
- output: number in A,
- exemplary value: $00001001\ 00000110_{UP\text{-}BCD} = 10010110_{P\text{-}BCD}$ (see Table 2.11).

Table 2.11 Implementation in Code and Intermediate Results

Code	Description	A	B
MOV A,#09h	Input a first number (tens)	00001001	
MOV B,#06h	Input a second number (unities)	00001001	00000110
SWAP A	Change nibbles $A_{7...4} \leftrightarrow A_{3...0}$	10010000	00000110
ADD A,B	Add B to A	10010110	00000110

```
 1       ;**********************************************************************
 2       ;* Conversion UP-BCD to P -BCD *
 3       ;**********************************************************************
 4       0009            n EQU 09h
 5       0006            m EQU 06h               ;{nm} 0906 UP-BCD = 96 DEC
 6
 7       0000: 74 09     MOV A,#n                ;input a first number
 8       0002: 75 F0 06  MOV B,#m                ;input a second number
 9       0005: 12 00 0A  LCALL UP_BCD_P_BCD
10
11       0008: 80 FE     STOP:                   SJMP STOP
12          ;----------------------------------------------------------------
13       000A:           UP_BCD_P_BCD:
14       000A: C4        SWAP A                  ;change nibbles A7..4<->A3..0
15       000B: 25 F0     ADD A,B                 ;add B to A
16                                               ;result in A
17       000D: 22        RET
18       ;--- end of file ---
```

2.2.12 Conversion UP-BCD to ASCII

- input: number in A – tens, B – unities,
- output: number in A – tens, B – unities,
- exemplary value: $00001001\ 00000110_{UP\text{-}BCD} = 00111001\ 00110$ 110_{ASCII} (see Table 2.12).

Table 2.12 Implementation in Code and Intermediate Results

Code	Description	A	B
MOV A,#09h	Input a first number	00001001	
MOV B,#06h	Input a second number	00001001	00000110
]ADD A,#30h	Add 30_{HEX} to A	00111001	00000110
XCH A,B	Exchange $A \leftrightarrow B$	00000110	00111001
ADD A,#30h	Add 30_{HEX} to A	00110110	00111001
XCH A,B	Exchange $A \leftrightarrow B$	00111001	00110110

```
1      ;*********************************************************************
2      ;*              Conversion UP-BCD -> ASCII*
3      ;*********************************************************************
4      0009            n EQU 09h
5      0006            m EQU 06h                    ;{nm} 0906 UP-BCD = 96 DEC
6
7      0000: 74 09     MOV A,#n                     ;input a first number
8      0002: 75 F0 06  MOV B,#m                     ;input a second number
9      0005: 12 00 0A  LCALL UP_BCD_ASCII
10                                                  ;result in A-tens, B-unities
11     0008: 80 FE     STOP:                        SJMP STOP
12             ;------------------------------------------------------------------------------
13     000A:           UP_BCD_ASCII:
14     000A: 24 30     ADD A,#30h                   ;add 30h to A
15     000C: C5 F0     XCH A,B                      ;exchange A<->B
16     000E: 24 30     ADD A,#30h                   ;add 30h to A
17     0010: C5 F0     XCH A,B                      ;exchange A<->B
18     0012: 22        RET
19     ;--- end of file ---
```

2.2.13 Conversion ASCII to BIN

- input: number in A – tens, B – unities,
- output: number in A,
- exemplary value: $00111001\ 00110110_{ASCII} = 01100000_{BIN}$ (see Table 2.13).

Table 2.13 Implementation in Code and Intermediate Results

Code	Description	A	B	R0
MOV A,#39h	Input a first number	00111001		
MOV B,#36h	Input a second number	00111001	00110110	
ANL B,#0Fh	Clear higher nibble $B_{7...4}$	00111001	00000110	
ANL A,#0Fh	Clear higher nibble $A_{7...4}$	00001001	00000110	
MOV R0,B	Make a copy to B	00001001	00000110	00000110
MOV B,#10	Let B be 10_{DEC}	00001001	00001010	00000110
MUL AB	Multiply A by B	01011010	00000000	00000110
ADD A,R0	Add R0 to A	01100000	00000000	00000110

```
1      ;*********************************************************************
2      ;*              Conversion ASCII to BIN    *
3      ;*********************************************************************
4      0039            n EQU 39H
```

```
 5    0036                m EQU 36H                ;{nm}=3936 ASCII=96 DEC
 6
 7    0000: 74 39         MOV A,#n                 ;input a first number
 8    0002: 75 F0 36      MOV B,#m                 ;input a second number
 9    0005: 12 00 0A      LCALL ASCII_BIN
10                                                 ;result in A
11    0008: 80 FE         STOP:                    SJMP STOP
12         ;-----------------------------------------------------------------------------
13    000A:               ASCII_BIN:
14    000A: 53 F0 0F      ANL B,#0Fh               ;clear higher nibble B7..4
15    000D: 54 0F         ANL A,#0Fh               ;clear higher nibble A7..4
16    000F: A8 F0         MOV R0,B                 ;make a copy to R0
17    0011: 75 F0 0A      MOV B,#10                ;let B be 10DEC
18    0014: A4            MUL AB                   ;multiply A by B
19    0015: 28            ADD A,R0                 ;add R0 to A
20    0016: 22            RET
21         ;--- end of file ---
```

2.2.14 Conversion ASCII to P-BCD

- input: number in A – tens, B – unities,
- output: number in A,
- exemplary value: $00110111\ 00111000_{ASCII} = 01111000_{P\text{-}BCD}$ (see Table 2.14).

Table 2.14 Implementation in Code and Intermediate Results

Code	Description	A	B
MOV A,#37h	Input a first number	00110111	
MOV B,#38h	Input a second number	00110111	00111000
ANL A,#0Fh	Clear higher nibble of $A_{7...4}$	00000111	00111000
SWAP A	Change nibbles $A_{7...4} \leftrightarrow A_{3...0}$	01110000	00111000
ANL B,#0Fh	Clear higher nibble of $B_{7...4}$	01110000	00001000
ADD A,B	Add B to A	01111000	00001000

```
 1    ;********************************************************************************
 2    ;*              Conversion ASCII to P-BCD *
 3    ;********************************************************************************
 4    0037                n EQU 37H
 5    0038                m EQU 38H                ;{nm} 3738 ASCII=78 DEC
 6
 7    0000: 74 37         MOV A,#n                 ;input a first number
 8    0002: 75 F0 38      MOV B,#m                 ;input a second number
 9    0005: 12 00 0A      LCALL ASCII_P_BCD
```

```
10                                          ;result in A
11    0008: 80 FE      STOP:                SJMP STOP
12       ;-----------------------------------------------------------------------
13    000A:            ASCII_P_BCD:
14    000A: 54 0F      ANL A,#0Fh           ;clear higher nibble of A7..4
15    000C: C4         SWAP A               ;change nibbles A7..4<->A3..0
16    000D: 53 F0 0F   ANL B,#0Fh           ;clear higher nibble of B7..4
17    0010: 25 F0      ADD A,B              ;add B to A
18    0012: 22         RET
19       ;--- end of file ---
```

2.2.15 Conversion ASCII to UP-BCD

- input: number in A – tens, B – unities,
- output: number in A – tens, B – unities,
- exemplary value: $00110111\ 00111000_{ASCII} = 00000111\ 00001000_{UP\text{-}BCD}$ (see Table 2.15).

Table 2.15 Implementation in Code and Intermediate Results

Code	Description	A	B
MOV A,#37h	Input a first number	00110111	
MOV B,#38h	Input a second number	00110111	00111000
ANL A,#0Fh	Clear higher nibble of A7...4	00000111	00111000
ANL B,#0Fh	Clear higher nibble of B7...4	00000111	00001000

```
1     ;********************************************************************
2     ;*              Conversion ASCII to UP-BCD *
3     ;********************************************************************
4     0037             n EQU 37H
5     0038             m EQU 38H             ;{nm} 3738 ASCII=78 DEC
6
7     0000: 74 37      MOV A,#n             ;input a first number
8     0002: 75 F0 38   MOV B,#m             ;input a second number
9     0005: 12 00 0A   LCALL ASCII_UP_BCD
10                                          ;result in A-tens, B-unities
11    0008: 80 FE      STOP:                SJMP STOP
12       ;-----------------------------------------------------------------------
13    000A:            ASCII_UP_BCD:
14    000A: 54 0F      ANL A,#0Fh           ;clear higher nibble of A7..4
15    000C: 53 F0 0F   ANL B,#0Fh           ;clear higher nibble of B7..4
16    000F: 22         RET
17       ;--- end of file ---
```

At the end of this chapter, we will present a subroutine for converting an ordinary fraction of the numerator/denominator form to a binary fraction. The algorithm checks if the numerator is less than the denominator, and if so, the conversion takes place. Otherwise, the OV flag is set.

2.2.16 Conversion BIN Fraction (num/denom) to BIN Fraction (dot notation)

- input: number in A – numerator, B – denominator,
- output: number in A – binary fraction in the form 0.xxx or void if numerator ≥ denominator and then OV = 1,
- exemplary value: 00000101_{BIN} / 00000110_{BIN} = $0.110101(01)_{BIN}$.

Table 2.16 shows the state of the registers after the first loop cycle, in which a bit with weight 2^{-1} is determined. Subsequent bits are determined according to the rule described in Example 2.6 presented in section 2.1.

Table 2.16 Implementation in Code and Intermediate Results

Code	Description	A	B	R1
MOV A,#5	Input numerator	00000101		
MOV B,#6	Input denominator	00000101	00000110	
MOV R2,#7	How many digits (precision)	00000101	00000110	
MOV R1,#0	Clear R1	00000101	00000110	00000000
LOOP: RL A	Rotate left	00001010	00000110	00000000
CLR C	Clear C flag	00001010	00000110	00000000
SUBB A,B	Compare A and B	00000100	00000110	00000000
JNC SKIP	Jump if not less	00000100	00000110	00000000
ADD A,B	Add B to A	---	---	---
SKIP: CPL C	Invert C flag	00000100	00000110	00000000
XCH A,R1	Exchange A ↔ R1	00000000	00000110	00000100
RLC A	Rotate left with carry bit	00000001	00000110	00000100
XCH A,R1	Exchange again	0000100	00000110	00000001
DJNZ R2,LOOP	Repeat for next digit	0000100	00000110	00000001
MOV A,R1	Copy a result to A	State after last lap: 01101010		

```
 1      ;*****************************************************************
 2      ;*              Conversion BIN fraction num/denom to dot notation*
 3      ;*****************************************************************
 4      0005            n EQU 5                 ;n=5 DEC
 5      0006            m EQU 6                 ;m=6 DEC
 6      0000: 74 05     MOV A,#n                ;input numerator
 7      0002: 75 F0 06  MOV B,#m                ;input denominator
 8      0005: 12 00 0A  LCALL FRACTION_BIN
 9                                              ;result in A
10      0008: 80 FE     STOP:           SJMP STOP
11      ;------------------------------------------------------------------------
12      000A:           FRACTION_BIN:
13      000A: F8        MOV R0,A                ;make a copy to R0
14      000B: C3        CLR C                   ;clear C flag
15      000C: 95 F0     SUBB A,B                ;check if num<denom
16      000E: E8        MOV A,R0                ;retrieve original value
17      000F: 40 03     JC LESS                 ;skip if num<denom
18      0011: D2 D2     SETB OV                 ;else set flag and stop
19      0013: 22        RET
20      0014:           LESS:
21      0014: 7A 07     MOV R2,#7               ;how many digits (precision)
22      0016: 79 00     MOV R1,#0               ;clear R1
23      0018:           LOOP:
24      0018: 23        RL A                    ;rotate left
25      0019: C3        CLR C                   ;clear C flag
26      001A: 95 F0     SUBB A,B                ;compare A and B
27      001C: 50 02     JNC SKIP                ;jump if not less
28      001E: 25 F0     ADD A,B
29      0020:           SKIP:
30      0020: B3        CPL C                   ;invert C
31      0021: C9        XCH A,R1                ;exchange A and R1
32      0022: 33        RLC A                   ;rotate with carry bit
33      0023: C9        XCH A,R1                ;exchange again
34      0024: DA F2     DJNZ R2,LOOP            ;repeat for next digit
35      0026: E9        MOV A,R1
36      0027: 22        RET
37      ;--- end of file ---
```

2.3 SIGNED NUMBERS

2.3.1 The Sign-magnitude Representation

The number consists of two fields: the sign '+'/'−' and the magnitude wherein '+' sign is very often omitted as the default value. This way of expressing the signed number is commonly used in usual human activities. We can find many examples, e.g.:

- trends in something (e.g. body weight, prices, demography, etc.);
- in science and engineering: measuring angles, outdoor thermometer (particularly winter season), acidity or alkalinity Ph;
- accounting and banking: looking at purchase bill (discounts and payable), account balance.

The value of a number in sign-magnitude (SM) notation for $p = 2$ is calculated from the following formula (2.2):

$$A = (-1)^{a_{n-1}} \cdot (a_{n-2} \cdot 2^{n-2} + \ldots + a_1 \cdot 2 + a_0 + a_{-1} \cdot 2^{-1} + \ldots + a_{-m} \cdot 2^{-m})$$
$$= (-1)^{a_{n-1}} \cdot \sum_{i=-m}^{n-2} a_i \cdot 2^i \tag{2.2}$$

The highest bit a_{n-1} is the sign bit, and the remaining bits form a magnitude of determined identically as in BIN format. The sign can be expressed by assigning '+' to '0' and '−' to '1', respectively. Above formula can be generalized for any value of the system base. However, the problem is how to encode the sign of the number. From the analysis of the formula it follows that if the highest digit is even, then the number is non-negative. Such a convention of writing the sign of a number, although formally correct, is not commonly used except in the case of $p = 2$. So, let's remain with this case. The use of the same symbols 0 and 1 to denote the sign of a number and the consecutive digits of the number facilitates the implementation of arithmetic operations on multibit numbers in the SM notation by a classical processor, which implements the principles of 1-bit arithmetic described in section 1.1. A particular property of sign-magnitude notation is the double representation of the number 0. For example, using 5 bits we +0 and –0:

$$+ 0_{DEC} = \underline{0}0000_{SM}, \quad -0_{DE} = \underline{1}0000_{SM}$$

For distinction, the sign bit in the book will be underlined. In other books, e.g. [Pochopień 2012], the sign bit is separated by a dot character, but this can lead to misinterpretation when trying to write numbers with fractional parts in SM!

Example 2.9: A number in the SM notation and its decimal equivalent DEC:

    a. $\underline{1}\,0101.11_{SM} = -5.75_{DEC}$
    b. $\underline{0}\,1100.01_{SM} = +12.25_{DEC}$

Exercise 2.6: Represent the given decimal number DEC in the SM notation:

    a. $+23.5_{DEC}$
    b. $+17.3_{DEC}$
    c. -11.25_{DEC}
    d. -1_{DEC}

The advantage of the SM notation is the simplicity of interpreting the number by the user. Unfortunately, this notation also has some disadvantages. Firstly, for $p \neq 2$ the SM format is not optimal because of using only $2*p^{n-1+m}$ of p^{n+m} possible combinations of digits. Another difficulty is related to realization of arithmetic operations. Since the ALU of the processor performs them on all the bits of the arguments stored in the registers, including the sign bits, this may lead to an incorrect result in some cases. What can be done in such a situation to fix the result? Before performing the operation, you need to clear the sign bits of both arguments, perform the operation and based on the sign bits and the information on type of operation determine the sign bit of the result.

However, the addition or subtraction of two 8-bit numbers with different signs is a little bit complicated. Hence another commonly used format for signed numbers is 2's complement. To understand the properties of this notation, we have to recall the theory of complements. The information presented below is valid for numbers with any base $p = 2, 3, ..., \infty$. It must be mentioned that there are also alternative ways of writing numbers with signs, such as the offset notation adopted by the IEEE society and published in its P-754 standard defining the floating-point format and operations and discussed in Chapters 4 and 5.

2.3.2 Complements – Theory and Its Usage

In mathematics and computing, the complements are efficient techniques to encode a symmetric range of positive and negative numbers. Two types of complements have gained widespread acceptance (2.3):

1. p complement of L

$$\overline{\overline{L}} = p^n - L \quad \text{for} \quad L \neq 0$$
$$\overline{\overline{L}} = 0 \quad\quad \text{for} \quad L = 0 \tag{2.3}$$

2. p – 1 complement of L

$\bar{L} = p^n - L - p^{-m}$, where n is the number of digits in the integer part and m is the fractional part, respectively.

Example 2.10: The p and p-1 complements:

a. $L = 823_{DEC} \rightarrow p = 10, n = 3, m = 0$ b. $L = 101.1_{BIN} \rightarrow p = 2, n = 3, m = 1$

$\bar{L} = 10^3 - 823 = 177_{DEC}$ $\bar{L} = 1000_{BIN} - 101.1_{BIN} = 010.1_{BIN}$

$\underbar{L} = 10^3 - 823 - 10^0 = 176_{DEC}$ $\underbar{L} = 1000_{BIN} - 101.1_{BIN} - 0.1_{BIN} = 010.0_{BIN}$

Properties of p and p-1 complements:

Property no. 1. $\bar{L} = p^n - L - p^{-m} = \bar{L} - p^{-m}$ $\rightarrow$ 1a. $\bar{L} = \underbar{L} + p^{-m}$

Property no. 2. $\bar{L} + L = 1\underbrace{00xxx0.0xxx0}_{n+m}$ $\rightarrow$ 2a. $\bar{L} = -L$, if carry bit on position n+m+1 is discarded

Example 2.11: Number added to its p complement:

a.

823_{DEC}
$+\ 177_{DEC}\ =\ L$
$\underbrace{1\,000}_{n+m\,DEC}\ =\ \bar{L}$

b.

101.1_{BIN}
$+\ 010.1_{BIN}\ =\ L$
$\underbrace{1\,000.0}_{n+m\ BIN}\ =\ \bar{L}$

Property no. 3. $\underbar{L} + L = \underbrace{(p-1)(p-1)...\,(p-1),\ (p-1)(p-1)...\,(p-1)}_{n+m}$

Example 2.12: Number added to its p-1 complement:

a.

823_{DEC}
$+\ 176_{DEC}\ =\ L$
$\underbrace{999}_{n+m\,DEC}\ =\ \underbar{L}$

b.

101.1_{BIN}
$+\ 010.0_{BIN}\ =\ L$
$\underbrace{111.1}_{n+m\ BIN}\ =\ \underbar{L}$

🕱 **REMEMBER!**

- $\bar{L}$ can be determined by adding 1 to the lowest digit of $\underbar{L}$.
- $\bar{L}$ can be treated as a number with the opposite sign to L if we discard 1 in the n+m+1 position.

- $\bar{L}$ is determined by subtracting individually each digit of L from the largest allowed digit at the given base, e.g. for $p = 2$ that means inversion of bits due to 2-bit = $\overline{bit}$.
- For $p = 2$ $\bar{L}$ it is denoted as a 1's complement and $\bar{\bar{L}}$ as a 2's complement.

Complements can be used, e.g., in the implementation of addition and subtraction of numbers coded in the SM notation, as will be shown in the chapter on arithmetic operations. Nevertheless, its primary usage is representation of signed numbers as also would be presented below.

2.3.3 The 2's Complement Representation

In this notation, abbreviated 2's in this book, the a_{n-1} bit is the sign bit, but unlike the SM notation, it additionally contributes with -2^{n-1} weighting factor to the value of the number. It can be then determined according to the formula (2.4):

$$A = -a_{n-1} \cdot 2^{n-1} + a_{n-2} \cdot 2^{n-2} + \dots + a_1 \cdot 2 + a_0 + a_{-1} \cdot 2^{-1} + \dots + a_{-m} \cdot 2^{-m}$$
$$= -a_{n-1} \cdot 2^{n-1} + \sum_{i=-m}^{n-2} a_i \cdot 2^i \qquad (2.4)$$

Example 2.13: Number in 2's complement notation and its DEC equivalent:

a. $101.0|_{U2} = -2^2 + 0 \cdot 2^1 + 1 \cdot 2^0 + 0 \cdot 2^{-1} + 1 \cdot 2^{-2} = -2.75_{DEC}$
b. $1101.0|_{U2} = -2^3 + 1 \cdot 2^2 + 0 \cdot 2^1 + 1 \cdot 2^0 + 0 \cdot 2^{-1} + 1 \cdot 2^{-2} = -2.75_{DEC}$
c. $11101.0|_{U2} = -2^4 + 1 \cdot 2^3 + 1 \cdot 2^2 + 0 \cdot 2^1 + 1 \cdot 2^0 + 0 \cdot 2^{-1} + 1 \cdot 2^{-2} = -2.75_{DEC}$
d. $0101.0|_{U2} = -0 \cdot 2^3 + 1 \cdot 2^2 + 0 \cdot 2^1 + 1 \cdot 2^0 + 0 \cdot 2^{-1} + 1 \cdot 2^{-2} = +5.25_{DEC}$
e. $00101.0|_{U2} = -0 \cdot 2^4 + 0 \cdot 2^3 + 1 \cdot 2^2 + 0 \cdot 2^1 + 1 \cdot 2^0 + 0 \cdot 2^{-1} + 1 \cdot 2^{-2} = +5.25_{DEC}$

The way of sign changing of a number in 2's notation is of great practical importance. It is enough to invert all bits and add a 1 to the lowest bit (first from the right).

Example 2.14: Changing the sign of a number in the 2's notation:

a.

$0110.|_{2's} \quad +6.5_{DEC}$
$\Downarrow$

```
  1001.0
+ 0000.1
```
$1001.|_{2's} \quad -6.5_{DEC}$

b.

$1001.|_{2's} \quad -6.5_{DEC}$
$\Downarrow$

```
  0110.0
+ 0000.1
```
$0110.|_{2's} \quad +6.5_{DEC}$

Another way to change the sign is to subtract a number from zero according to the observation that $-X = 0 - X$, where the minus on the left represents the number with the opposite sign to X, and the second minus represents the subtraction operation. Be aware, that it looks the same but it is not the same for mathematicians!

Example 2.15: Changing the sign of a number in the 2's notation by subtracting from zero:

a.

	0000.0	0.0$_{DEC}$	and		0000.0	0.0$_{DEC}$
−	0110.1	+ 6.5$_{DEC}$		−	1001.1	− 6.5$_{DEC}$
	1001.1	− 6.5$_{DEC}$			0110.1	+ 6.5$_{DEC}$

b.

🕯️ **REMEMBER!**

- Duplication of the highest bit (first from the left) does not change the value of the number in 2's notation. It is called the sign extension.
- The highest bit informs about the sign of the number, '0' is for non-negative number and '1' for negative one, but it also affects the value of the number.
- Typical way of sign change of 2's number is inverting all bits and adding the 1 to the lowest bit (first from the right).

Exercise 2.7: Represent the assumed DEC number in 2's notation:

  a. +3.125$_{DEC}$
  b. −17.5$_{DEC}$
  c. −1$_{DEC}$
  d. +1$_{DEC}$

Number with fraction in 2's notation can be represented easily as a complement to a greater (absolutely) number. Thus, you can apply the conversion method learned in section 2.1 only to the positive fractional part of this number as shown here: $-2/9 = -1 + 7/9$ or $-12/5 = -3 + 3/5$.

2.4 CONVERSIONS AND CHANGE OF SIGN

The chapter will show how to programmatically convert numbers with sign between 2's and SM formats, and how to change the sign of numbers in 2's notation. Please note the number in 2's complement format has no separate representation for positive and negative zero.

2.4.1 Change of Sign for 2's Number

- input: number in A,
- output: number in A,
- exemplary value: $11101111_{2's} \rightarrow 00010001_{2's}$ (see Table 2.17).

Table 2.17 Implementation in Code and Intermediate Results

Code	Description	A
MOV A,# 11101111b	Input -17_{DEC} as 2's	11101111
XRL A,#0FFH	Invert A	00010000
INC A	Increment A by 1	00010001

```
 1      ;********************************************************************
 2      ;*                  Conversion of sign for 2's number    *
 3      ;********************************************************************
 4      ;n EQU 17           ;n=+17 DEC
 5      00EF                n EQU 11101111B              ;n= -17 DEC
 6
 7      0000: 74 EF         MOV A,#n                     ;input a number
 8      0002: 12 00 07      LCALL _2s_2s
 9                                                       ;result in A
10      0005: 80 FE         STOP:            SJMP STOP
11         ;----------------------------------------------------------------
12      0007:               _2s_2s:
13      0007: 64 FF         XRL A,#0FFH                  ;invert A
14      0009: 04            INC A                        ;increase A by 1
15      000A: 22            RET
16      ;--- end of file ---
```

2.4.2 Conversion SM to 2's Notation

- input: number in A,
- output: number in A,
- exemplary value: $10000111_{SM} \rightarrow 11111001_{2's}$ (see Table 2.18).

Table 2.18 Implementation in Code and Intermediate Results

Code	Description	A
MOV A,# 10000111b	Input a number -7_{DEC} as SM	10000111
JNB ACC.7,SKIP	Skip if positive	
XRL A,#7FH	Invert bits $A_{6...0}$	11111000
INC A	Increase A by 1	11111001
SKIP:		11111001

```
 1      ;***********************************************************************
 2      ;*                       Conversion SM to 2's      *
 3      ;***********************************************************************
 4      0087                     n EQU 10000111B           ;n=−7DEC
 5
 6      0000: 74 87              MOV A,#n                  ;input a number
 7      0002: 12 00 07           LCALL SM_2s
 8                                                         ;result in A
 9      0005: 80 FE              STOP:          SJMP STOP
10      ;----------------------------------------------------------------------
11      0007:                    SM_2s:
12      0007: 30 E7 03           JNB ACC.7,SKIP            ;if positive
13      000A: 64 7F              XRL A,#7FH                ;invert bits A6..0
14      000C: 04                 INC A                     ;increase A by 1
15      000D:                    SKIP:
16      000D: 22                 RET
17      ;--- end of file ---
```

2.4.3 Conversion 2's Notation to SM

- input: number in A,
- output: number in A,
- exemplary value: 11111001_{U2} → 10000111_{SM}.

The algorithm returns a correct result for an input number between -127_{DEC} and $+127_{DEC}$ (see Table 2.19).

Table 2.19 Implementation in Code and Intermediate Results

Code	Description	A
MOV A,# 11111001b	Input a number -7_{DEC} as 2's	11111001
JNB ACC.7,SKIP	Skip if positive	11111001
XRL A,#7FH	Invert bits $A_{6...0}$	10000110
INC A	Increase A by 1	10000111
SKIP:		10000111

Please note that we get the identical algorithm as for SM to 2's complement conversion!

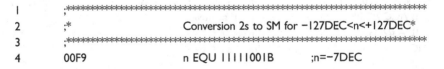

```
 1      ;*************************************************************************
 2      ;*               Conversion 2s to SM for −127DEC<n<+127DEC*
 3      ;*************************************************************************
 4      00F9                     n EQU 11111001B           ;n=−7DEC
```

```
 5
 6      0000: 74 F9          MOV A,#n                ;input a number
 7      0002: 12 00 07       LCALL _2s_SM
 8                                                   ;result in A
 9      0005: 80 FE          STOP:       SJMP STOP
10         ;-------------------------------------------------------------------------------
11      0007:                _2s_SM:
12      0007: 30 E7 03       JNB ACC.7,SKIP          ;skip if positive
13      000A: 64 7F          XRL A,#7FH              ;invert bits A6..0
14      000C: 04             INC A                   ;increase A by 1
15      000D:                SKIP:
16      000D: 22             RET
17      ;--- end of file ---
```

Chapter 3

Basic Arithmetic on Fixed-point Numbers

3.1 OPERATIONS ON UNSIGNED NUMBERS

3.1.1 Working with Natural Binary Code

As was stated previously in section 2.1, let us recall that any number is represented on n+m bits, where n is the number bits of the integer part, and m is the number bits of the fractional part.

Addition of two $(n + m)$-bit BIN numbers returns the result on $(n + m + 1)$ bits. It can be seen that for any base p this extra bit can take the value 0 or 1. Reserving bits for the fractional part is limiting the range in the integer part of the number. Appendix A gives the smallest and largest value of a number for given n and m, assuming $n + m = 8$ and $n + m = 16$ and for various position of point. This information will allow the reader to find out in what numerical range arithmetic operations can be performed. The length of the word in the 8051 processor is fixed – 8 bits, so it can store 256 different combinations. The $n + m$ cannot exceed 8 bits. When considering numbers with fractional part, one should make sure that both numbers contain the same number of m digits in the fractional part. Then the position of the point in the result is identical to that in the input arguments, although the microprocessor is 'not aware' the existence and position of the comma. In many programming languages the fixed-point format is dedicated for integer type of variables without fractional part (see Appendix B). The numbers with fractions are mostly represented in floating-point format realized in hardware like floating-point unit (FPU) or programmatically by dedicated math library. We can also realize real numbers in fixed-point format. Here a priori some bits are reserved for integer and other for fractional part. For example, if assumed 1-byte-long number, the notation 1.7 means 1 bit for integer part and 7 bits after binary point. This convention is supported by, e.g., some of Microchip's AVR architecture processors through the FMUL, FMULS and FMULSU instructions that allow multiplication of two numbers with an unsigned fraction, two signed numbers and an unsigned with signed numbers. The prefix F related to fraction just indicates this functionality. No additional shifts of the multiplication result or another way of point positioning are

DOI: 10.1201/9781003363286-3

needed. Another of few example is so called 32-bit _IQ format introduced by Texas Instruments for fixed-point numbers with fractional part. At the moment of declaration of variable type, its precision and the numerical range are explicitly forced, depending on the number of bits for the fractional part. Thus, the notation _IQ30 means that 30 bits of 32 available bits are used to express the fraction and the remaining 2 bits for the integer part. Using the 2's convention, this gives the range [–2;1.999999999] with a resolution (precision) of 0.000000001. All combinations from _IQ0 to _IQ31 are allowed, so paradoxically one can speak of a fixed-point format with a floating point is permitted! Another example of IQ format application can be using it to express numbers during performing CORDIC algorithm suited for estimation of nonlinear functions. The ST Microelectronics company prepared STM32G4-CORDIC co-processor. It provides hardware acceleration of some mathematical functions, notably trigonometric, commonly used in motor control, metering, signal processing and many other applications. It speeds up the calculation of these functions compared to a software implementation, freeing up processor cycles in order to perform other tasks. In this case, the q1.31 or q1.5 formats are available. For processors without hardware support as FPU or CORDIC units this mathematical capability can be realized by software CORDIC library or math library.

Let's go back to 8051 CPU and way how the addition of 2 bytes can be done with instruction ADD:

$$\text{ADD A,\#data} \qquad\qquad \{C,A\} \leftarrow A + data$$

The result of addition of each pair of bits is calculated according to the rules given in section 1.1, considering the carry-over from the previous bit position. The addition starts from the youngest bits, i.e. customarily the right-hand side. If the ADD instruction results in $C = 1$, it means that the BIN range is exceeded for the assumed values of n and m. The result should be discarded. However, the correct result is obtained by treating C as an additional $n + 1$ bit of the result (see Example 3.1a).

Example 3.1: Addition of BIN numbers:

a. n = 5, m = 3			b. n = 5, m = 3		
	11111.010_{BIN}	$=31.25_{DEC}$		10011.010_{BIN}	$=19.25_{DEC}$
+	00001.000_{BIN}	+ $=1.00_{DEC}$	+	00001.011_{BIN}	+ $=1.375_{DEC}$
$C = 1$	00000.010_{BIN}	$=32.25_{DEC}$	$C = 0$	10100.101_{BIN}	$=20.625_{DEC}$

Implementation in code for case (a):

```
MOV  A,#11111010B  ;first number
ADD  A,#00001000B  ;add the second number to the first one
                   ;store a result in the A and C
```

The addition of two double-byte (n + m) = 16 BIN numbers is performed in two steps. First add the lower bytes with the ADD instruction, and then the higher ones with the ADDC instruction, which considers the C flag from the previous addition. The double-byte width is sometimes called 'word' format.

Example 3.2: Addition of two double-byte BIN numbers:

a. n = 8, m = 8   b. n = 14, m = 2

```
        |                                      |
    00011001.11110101BIN              10011001 111101.01BIN
 +  10011101.10000101BIN           +  11011101100001.01BIN
 _____       _____
 0   10110111.01111010BIN           1   01101110111110.10BIN
```

Implementation for case of (a):

```
MOV   A,#11110101B    ;lower byte of the first number
ADD   A,#10000101B    ;add a lower byte of the second number
MOV   B,A             ;lower byte of the result in the B
MOV   A,#00011001B    ;higher byte of the first number
ADDC  A,#10011101B    ;add a higher byte of the second number
                      ;and store a result in the A and C
```

Exercise 3.1: Perform addition of BIN numbers:

a.   b.

```
    11.011101BIN                    1101.011BIN
 +  00.101101BIN                 +  1010.1100BIN
 _____            _____
 ?   ??.??????BIN                ?   ????.????BIN
```

Subtraction of two (n + m)-bit BIN numbers returns the result on (n + m + 1) bits, where, as previously, n is the number of bits of the integer part, m is the number of bits of the fractional part. It can be seen that for any base, p is the borrow flag (in practice it is the same C flag used during addition – its meaning depends on the operation context), which can take only the value 0 or 1. When considering numbers with fractional part, one should make sure that both numbers contain the same number of digits m in the fractional part. Then the position of the comma in the result is identical to that in the input arguments. The result of subtraction of each pair of bits is calculated according to the rules given in chapter 1.1, considering the borrow from the previous bit position. The operation starts from the lowest bits, i.e. customarily the right-hand side.

Subtracting two 1-byte numbers by 8051 CPU is done by the following pair of instructions:

```
CLR   C
SUBB  A,#data                    {C,A}←A-data-0=A-data
```

Before subtracting 1-byte two numbers, hence(n + m = 8), it is necessary to clear C flag to zero, preventing eventually wrong result caused by C = 1 flag by one of the previous executed instructions. Obtaining C = 1 after subtraction indicates a negative result that cannot be expressed in BIN notation. Alternatively, this state can be interpreted as borrow taken from the extra n + 1 bit of the first number. Then the result is correct in the BIN sense, since subtraction of a smaller number from a larger number has been performed.

Example 3.3: Subtraction of BIN numbers:

a. n = 2, m = 6

$$10.010100_{BIN}$$
$$-\quad 01.101001_{BIN}$$

$C = 0$ $\quad 00.101011_{BIN}$

b. n = 8, m = 0

$$00010100_{BIN}$$
$$-\quad 00101011_{BIN}$$

$C = 1$ $\quad 11101001_{BIN}$

Implementation for case of (a):

```
MOV  A,#10010100B  ;first number
CLR  C             ;clear C flag
SUBB A,#01101001B  ;subtract second number from the first one
                   ;store a result in the A and C
```

The subtraction of two double-byte (n + m) = 16 BIN numbers is performed in two steps. First subtract the lower bytes by the SUBB instruction with cleared the C flag before, and then subtract the higher ones again with SUBB, which considers the C flag from the previous operation that is working as borrow bit.

Example 3.4: Subtraction of two double-byte BIN numbers:

a. n = 16, m = 0

$$\overset{0}{01011001}\quad 10011011_{BIN}$$
$$-\quad 00011101\quad 01110010_{BIN}$$

$C = 0$ $\quad 00111100\quad 00101001_{BIN}$

b. n = 3, m = 13:

$$\overset{1}{010.11001}\quad 00011011_{BIN}$$
$$-\quad 100.11101\quad 01110010_{BIN}$$

$C = 1$ $101.11011\quad 10101001_{BIN}$

Subtraction of two double-byte BIN numbers by 8051 CPU is very similar to the addition. The difference is that the ADD and ADDC instructions must be replaced by pair of SUBBs.

Implementation in code of case (a):

```
MOV    A,#10011011B    ;lower byte of the first number
CLR    C               ;clear C flag
SUBB   A,#01110010B    ;subtract the lower byte of the second number
MOV    B,A             ;store the lower byte of the result in the B
MOV    A,#01011001B    ;higher byte of the first number
SUBB   A,#00011101B    ;subtract the higher byte of the second number
                       ;and store the higher byte of the result in the A
```

Exercise 3.2: Perform subtraction of BIN numbers:

a.

$$11.011100_{BIN}$$
$$-\ \ 01.10101_{BIN}$$
$$\overline{?\quad ??.??????_{BIN}}$$

b.

$$0101.101\,1_{BIN}$$
$$-\ \ 1010.1101_{BIN}$$
$$\overline{?\quad ????.????_{BIN}}$$

Subtraction can also be done in another way, i.e. using addition and properties of complements.

Subtraction of two BIN numbers by means of complements requires substitution of the second number by its complement to p or p − 1. The result of the subtraction is the result of the addition or its complement, depending on the value of the carry bit. Details on the use of additions are given in the formulas below. The way how to compute the complement to p and p − 1 was given earlier in subchapter 2.3.

Using p complements for subtraction:

$$A - B = +(A + \bar{\bar{B}}) \quad \text{for} \ \ C \neq 0$$
$$A - B = -\overline{(A + \bar{\bar{B}})} \quad \text{for} \ \ C = 0$$

Using p − 1 complements for subtraction:

$$A - B = +(A + \bar{B} + 1 \cdot p^{-m}) \quad \text{for} \ \ C \neq 0$$
$$A - B = -\overline{(A + \bar{B})} \quad \text{for} \ \ C = 0$$

Example 3.5: Subtraction of two BIN numbers by means of p and p − 1 complements for p = 2:

$$A = 29_{DEC}$$
$$A = 011101_{BIN}$$
$$\bar{A} = 100010$$
$$\bar{\bar{A}} = 100011$$

$$B = 38_{DEC}$$
$$B = 100110_{BIN}$$
$$\bar{B} = 011001$$
$$\bar{\bar{B}} = 011010$$

a. A − B by means of $\bar{\bar{B}}$

	011101	A
+	011010	$\bar{\bar{B}}$
C = 0	110111	A + $\bar{\bar{B}}$

⇓

$$'-' \equiv 1 \quad 001001 \quad -\overline{(A + \bar{\bar{B}})}$$

b. B − A by means of $\bar{\bar{A}}$

	100110	B
+	100011	$\bar{\bar{A}}$
C = 1	001001	B + $\bar{\bar{A}}$

⇓

$$'+' \equiv 0 \quad 001001 \quad +(B + \bar{\bar{A}})$$

Implementation in code:

a.

```
MOV     A,#00011101B
MOV     B,#00100110B
XCH     A,B
CPL     A
INC     A
ADD     A,B
JC      end
CPL     A
INC     A
end:
CPL     C
```

b. would you try to continue?

```
MOV     A,#0001110
MOV     B,#00100110B
        ...
```

;sign of the result
;the result stored in the A

c. A − B by means of $\bar{B}$

	011101	A
+	011001	$\bar{B}$
C = 0	110110	$A + \bar{B}$
	⇓	
' − ' ≡ 1	001001	$-(\overline{A + \bar{B}})$

d. B − A by means of $\bar{A}$

	100110	B
+	100010	$\bar{A}$
C = 1	001000	$B + \bar{A}$
+	000001	p^{-m}
' + ' ≡ 0	001001	$+(B + \bar{A} + 1 \cdot p^{-m})$

Implementation in code:

c.

```
MOV  A,#00011101B
MOV  B,#00100110B
XCH  A,B
CPL  A
ADD  A,B
JC   skip
CPL  A
SJMP end
skip:
INC  A
end:
CPL  C
```

d. ?

```
MOV     A,#00011101B
MOV     B,#00100110B
        ...
```

;sign of the result
;result in A

Example 3.6: Subtraction by p and p − 1 complements for p = 10:

$A = 29_{DEC}$ $B = 38_{DEC}$

$\bar{A} = 70$ $\bar{B} = 61$

$\bar{\bar{A}} = 71$ $\bar{\bar{B}} = 62$

a. A − B by using $\bar{\bar{B}}$

	29	A
+	62	$\bar{\bar{B}}$
C = 0	91	$A + \bar{\bar{B}}$

⇓

' − ' ≡ 1 09 $-(\overline{\overline{A + \bar{\bar{B}}}})$

b. B − A by using $\bar{\bar{A}}$

	38	B
+	71	$\bar{\bar{A}}$
C = 1	09	$B + \bar{\bar{A}}$

⇓

' + ' ≡ 0 09 $+(B + \bar{\bar{A}})$

c. A − B by using $\bar{B}$

	29	A
+	61	$\bar{B}$
C = 0	90	$A + \bar{B}$

⇓

' − ' ≡ 1 09 $-(\overline{A + \bar{B}})$

d. B − A by using $\bar{A}$

	38	B
+	70	$\bar{A}$
C = 1	08	$B + \bar{A}$
+	01	p^{-m}

'+' ≡ 0 09 $+(B + \bar{A} + p^{-m})$

where m = 0

💀 REMEMBER!

- The sign bit of a result of subtraction done with complements can be determined by inversion of carry bit.
- The rules of applying complements can be used to any system base.

Exercise 3.3: Perform subtraction of BIN numbers using p and p − 1 complements:

    a. A = 42.5_{DEC} B = 68_{DEC} p = 2
    b. A = 75_{DEC} B = 13_{DEC} p = 10

Multiplication of two (n + m) –bits BIN numbers returns on (2n + 2m) bits. For general case of multiplication, i.e. A1*A2, where A1 is on {n1,m1} bits and A2 is on {n2,m2} bits, the result is on (n1 + n2, m1 + m2) bits. In 8051 processor, the multiplication is done by MUL instruction:

MUL AB $\{B_{15...8} A_{7...0}\}$ ← A*B ; 16-bits results

The classical 'on paper' multiplication method taught in primary school is easy to understand and can be realized by adding shifted results of multiplication of individual bits of the multiplied by all bits of multiplicand. This rule is illustrated in Example 3.7.

Example 3.7: Multiplication of two BIN numbers:

a.

$$10.10_{BIN}$$
$$*01.0_{BIN}$$

1010	
0000	
1010	
+ 0000	

$$0011.0010_{BIN}$$

b.

$$1.111_{BIN}$$
$$*110._{BIN}$$

1111	
0000	
1111	
+ 1111	

$$1100.001_{BIN}$$

The multiplication of the first number (multiplicand) by the bit of multiplier bit with a value of 0 results 0 ... 0 due to observation $0*x = 0$. Therefore, this component of the result can be omitted in the final addition. However, the next non-zero component of the sum should be shifted twice, not once.

Implementation in code for case of (a):

```
MOV        A,#00010010B
MOV        B,#0000101B
MUL        AB                    ;result is in B – higher byte, A – lower byte
```

The algorithm can be automated, presenting it in a form suitable for implementation in code of microcontrollers suffering from lack of multiplication instruction. An example of such a processor is the ATtiny family of Atmel Corporation, acquired in 2016 by Microchip Technology or 68HC08 chip delivered by Freescale Semiconductor. The rule of operation can be described as below:

1. Clear the result and the carry bit.
2. Copy the multiplier to the lower part of the result.
3. Shift to the right the result and carry bit becomes the highest bit of the result.
4. If the lowest bit of the result that is lost after shifting was set, add the multiplicand (first number) to the higher part of the result and store a carry-over after adding in the carry bit.
5. Repeat from step 3 for all multiplier bits; the number of laps depends on the number of multiplier bits.

Details of the algorithm are presented in Example 3.8, in which the previous lowest bit of result is underlined, while the carry bit is to the left of the higher part of the result. It also becomes the highest bit of the result after shifting right.

Example 3.8: Multiplication of two BIN numbers (another version):

a)   b)

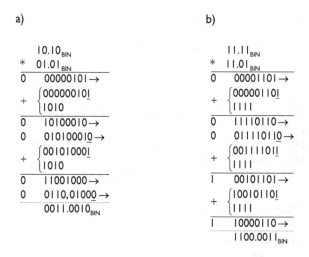

Exercise 3.4: Perform multiplication of BIN numbers:

a.   b.

$$
\begin{array}{r}
11.11_{BIN} \\
* \quad 0.101_{BIN} \\
\hline
???.?????_{BIN}
\end{array}
$$

$$
\begin{array}{r}
1.00_{BIN} \\
* \quad 0110_{BIN} \\
\hline
?????.???_{BIN}
\end{array}
$$

Multiplication of two double-byte (n + m = 16) BIN numbers is done according to the same principle. The bytes are multiplied instead of the individual bits and the resulting partial products are then added.

Example 3.9: Multiplication of double-byte BIN numbers:

```
          23  48 HEX
  *       AB  9C HEX
          2B  E0
      15  54
      30  18
  + 17 61
  ─────────────────
   17  A6  97  E0 HEX
```

Implementation in code:

- input: R1 – higher byte of multiplicand, R0 – lower byte of multiplicand, R3 – higher byte of multiplier, R2 – lower byte of multiplier,

- output: {R3R2R1R0} R3 – highest byte of result, R0 – lowest byte of result,
- exemplary value: $2348_{HEX} * AB9C_{HEX}$.

```
 1      ;******************************************************************
 2      ;* Multiplication of BIN numbers 2 bytes x 2 bytes *
 3      ;******************************************************************
 4      0000: 79 23        MOV R1,#23h
 5      0002: 78 48        MOV R0,#48h            ;multiplicand in {R1,R0}=2348HEX
 6      0004: 7B AB        MOV R3,#0ABh
 7      0006: 7A 9C        MOV R2,#9Ch            ;multiplier in {R3,R2}=AB9CHEX
 8      0008: 12 00 0D     LCALL MUL_BIN16X16
 9                                                ;result in {R3,R2,R1,R0}
10
11      000B: 80 FE        STOP: SJMP STOP
12      ;----------------------------------------------------------------
13      000D:              MUL_BIN16X16:
14      000D: E8           MOV A,R0
15      000E: 8A F0        MOV B,R2
16      0010: A4           MUL AB
17      0011: AD F0        MOV R5,B
18      0013: FC           MOV R4,A               ;R5*R4=R0*R2
19      0014: E9           MOV A,R1
20      0015: 8A F0        MOV B,R2
21      0017: A4           MUL AB                 ;B*A=R2*R1
22      0018: 2D           ADD A,R5
23      0019: FD           MOV R5,A
24      001A: E4           CLR A
25      001B: 35 F0        ADDC A,B
26      001D: FE           MOV R6,A               ;{R6R5R4} BANK 0
27      001E: E8           MOV A,R0
28      001F: 8B F0        MOV B,R3
29
30      0021: D2 D3        SETB RS0               ;BANK 1
31      0023: A4           MUL AB
32      0024: AD F0        MOV R5,B
33      0026: FC           MOV R4,A               ;R5*R4=R0*R2
34
35      0027: C2 D3        CLR RS0                ;BANK 0
36      0029: E9           MOV A,R1
37      002A: 8B F0        MOV B,R3
38      002C: D2 D3        SETB RS0               ;BANK 1
```

39	002E: A4	MUL AB	;B*A=R2*R1
40	002F: 2D	ADD A,R5	
41	0030: FD	MOV R5,A	
42	0031: E4	CLR A	
43	0032: 35 F0	ADDC A,B	
44			
45	0034: FE	MOV R6,A	;{R6R5R4}
46	0035: C2 D3	CLR RS0	;BANK 0
47	0037: AB 0E	MOV R3,6+8H	
48	0039: AA 0D	MOV R2,5+8H	
49	003B: A9 0C	MOV R1,4+8H	
50	003D: A8 04	MOV R0,4H	
51	003F: E9	MOV A,R1	
52	0040: 2D	ADD A,R5	
53	0041: F9	MOV R1,A	
54	0042: EA	MOV A,R2	
55	0043: 3E	ADDC A,R6	
56	0044: FA	MOV R2,A	
57	0045: E4	CLR A	
58	0046: 3B	ADDC A,R3	
59	0047: FB	MOV R3,A	
60	0048: 22	RET	
61	;--- end of file ---		

Division of two $(n + m)$-bits BIN numbers can return a result over infinitely many bits, e.g. $101_{BIN}/110_{BIN} = 0.1101(01)_{BIN}$. The limited word length of the microprocessor imposes truncating the fractional part of the result. Alternatively, the result of division can be represented as a quotient (integer part) and the remainder (rest of the division). There are many algorithms, mainly varying in complexity. The simplest one is based on the observation that division is equivalent to many repeated subtractions. In this algorithm, the divisor is subtracted from the so-called partial remainder. Initially, the divisor should be taken as the partial remainder. If the result of subtraction is non-negative, the quotient is increased by 1, and the current result of subtraction is taken as the partial remainder. Continue doing this until the remainder is less than the divisor, as indicated by the borrow bit being set. In practice, the carry bit C is playing that role. The final remainder of division is equal to the last result of subtraction, after which C = 0.

Example 3.10: Perform division of two BIN numbers using the consecutive subtraction method:

a.

111_{BIN} – divident
$- \ 010_{BIN}$ – divisor
$\overline{101}$ $C = 0 \Rightarrow$ quotient= 1_{DEC}
$- \ 010$
$\overline{011}$ $C = 0 \Rightarrow$ quotient= 2_{DEC}
$- \ 010$
$\overline{001}$ $C = 0 \Rightarrow$ quotient= $3_{DEC} = 011_{BIN}$
$- \ 010$
$\overline{111}$ $C = 1 \Rightarrow$ remainder= 001_{BIN}

b.

100_{BIN} – divident
$- \ 010_{BIN}$ – divisor
$\overline{010}$ $C = 0 \Rightarrow$ quotient= 1_{DEC}
$- \ 010$
$\overline{000}$ $C = 0 \Rightarrow$ quotient= $2_{DEC} = 010_{BIN}$
$- \ 010$
$\overline{110}$ $C = 1 \Rightarrow$ remainder= 000_{BIN}

The disadvantage of this method is the variable execution time, which depends on the dividend to divisor ratio. The greater the ratio, the longer it takes. There are more efficient ways of dividing, e.g. the differential method, shown in Example 3.11. This time the number of operations does not depend on the dividend to divisor ratio, but on the number of bits of the divisor. It is also worth mentioning about the comparison method, sometimes called the non-restitution method, described in detail, among others, in [Pochopień 2012].

Example 3.11: Division of two BIN numbers by differential method:

a. smaller by a larger number:

$$
110.1_{BIN} : 1000_{BIN} \rightarrow \quad
\begin{array}{l}
0.1101_{BIN} \ \text{–fraction} \\
\hline
1101_{BIN} \qquad : 10000_{BIN} \\
11010 \\
-10000 \\
\hline
10100 \\
- \ 10000 \\
\hline
10000 \\
- \ 10000 \\
\hline
00000_{BIN} \ \text{–reminder}
\end{array}
$$

Alternatively with reminder

$$
\begin{array}{l}
0_{BIN} \ \text{–quotient} \\
\hline
1101_{BIN} \quad : 10000_{BIN} \\
\hline
10000_{BIN} \ \text{–reminder}
\end{array}
$$

b. larger by a smaller number:

$$
\begin{array}{ll}
101.1_{BIN} & \text{—quotient and fraction} \\
\hline
10110_{BIN} \quad : 100_{BIN} & \\
-100 & \\
\hline
110 & \\
-100 & \\
\hline
100 & \\
-100 & \\
\hline
000_{BIN} \quad \text{—reminder} &
\end{array}
$$

or with reminder

$$
\begin{array}{ll}
101_{BIN} & \text{—quotient} \\
\hline
10110_{BIN} \quad : 100_{BIN} & \\
-100 & \\
\hline
110 & \\
- \quad 100 & \\
\hline
010_{BIN} \quad \text{—reminder} &
\end{array}
$$

Exercise 3.5: Perform division of number 110010_{BIN} by 1101_{BIN} using:

a. consecutive subtraction method,
b. differential method.

 INTERESTING FACTS!

If you start your experience with processors created by Atmel (currently Microchip), you may be surprised by the lack of division (and sometimes multiplication) operations in the instruction list of some microcontrollers of AVR family, e.g. ATtiny! In such case, the missing instruction should be replaced with a proper subroutine. You can find appropriate algorithms on the Microchip website.

Dividing two single-byte numbers by the 8051 CPU is very simple, because we have the DIV instruction:

DIV AB {A – quotient, B – remainder} ←A÷B; 16-bit result

Implementation in code:

MOV A,#00010110B

MOV B,#00000100B

DIV AB; result in A – quotient, B – remainder

Of course, for zero value of the divisor the processor cannot perform the division. Instead of engaging extra bit for this purpose in processor resources, the designers of the microprocessor architecture have used the OV flag to signal such a case (it concerns for 8051 CPU at least). This is another use of this flag besides informing about exceeding the range of numbers in 2's complement format.

Division of double-byte (n + m = 16) BIN number by (n + m = 8) number is not as simple as for single-byte arguments, because the DIV instruction cannot be used in this case. The solution is provided by the implementation of, e.g., the multiple subtraction method for double-byte arguments, which works according to the principle presented in Example 3.10. If the divisor is equal to 0, the algorithm does not perform any operations except setting the OV flag.

Implementation in code:

- input: B – higher byte of dividend, A – lower byte of dividend, R0 – divisor,
- output: R3 – higher byte of quotient, R2 – lower byte of quotient, B – higher byte of reminder, A – lower byte of reminder, OV – signal an attempt to divide by zero,
- exemplary value: $4487_{DEC}/100_{DEC}$.

```
 I     ;*************************************************************
 2     ;* Division of BIN numbers 2 bytes/I byte *
 3     ;* consecutive subtraction method *
 4     ;*************************************************************
 5     0011            n EQU 17
 6     0087            m EQU 135          ;17*256+135=4487 DEC
 7     0064            y EQU 100          ;y=100 DEC
 8
 9     0000: 75 F0 11  MOV B,#n           ;higher byte of dividend
10     0003: 74 87     MOV A,#m           ;lower byte of dividend
11     0005: 78 64     MOV R0,#y          ;divisor
12     0007: 12 00 0C  LCALL DIV_BIN8BY8
13                                        ;result in {R3R2}-quotient,
14                                        ;{BA}-reminder
15     000A: 80 FE     STOP: SJMP STOP
16     ;-----------------------------------------------------------
17     000C:           DIV_BIN8BY8:
18     000C: B8 00 03  CJNE R0,#0,LOOP
19     000F: D2 D2     SETB OV
20     0011: 22        RET
21     0012:           LOOP:
22     0012: C3        CLR C
```

23	0013: F9	MOV R1,A	
24	0014: 98	SUBB A,R0	
25	0015: C0 E0	PUSH ACC	
26	0017: E5 F0	MOV A,B	
27	0019: 94 00	SUBB A,#0	
28	001B: C0 E0	PUSH ACC	
29	001D: 40 0D	JC LESS	
30	001F: EA	MOV A,R2	
31	0020: 24 01	ADD A,#1	
32	0022: FA	MOV R2,A	
33	0023: 50 01	JNC LESSTHAN256	
34	0025: 0B	INC R3	
35	0026:	LESSTHAN256:	
36	0026: D0 F0	POP B	;reminder
37	0028: D0 E0	POP ACC	;reminder
38	002A: 80 E6	SJMP LOOP	
39	002C:	LESS:	
40	002C: D0 E0	POP ACC	
41	002E: D0 E0	POP ACC	
42	0030: E9	MOV A,R1	
43	0031: 22	RET	
44	;--- end of file ---		

In situations where the execution time of the division is a critical factor, it is proposed to use the differential method in a version suitable for easy implementation in code. We will name it the 'differential method II'. Its operation can be characterized as follows, assuming zero values of quotient and partial remainder at start:

1. Put the dividend to the right after the rest.
2. Shift the partial remainder to the left together with dividend.
3. Subtract the divisor from the partial remainder and take the result as the new value of partial remainder.
4a. If the result of the subtraction is negative, add the divisor back to the partial remainder to get its original value, the lowest quotient bit is 0.
4b. If the result of the subtraction is non-negative, the lowest quotient bit is 1.
5. Shift the quotient to the left.
6. Repeat from step 2 for all the bits of the divisor.

The above rules of operation are used in Example 3.12.

Example 3.12: Division of BIN numbers by differential method II:

a) $1010_{BIN}:011_{BIN}$ b) $1110_{BIN}:101_{BIN}$

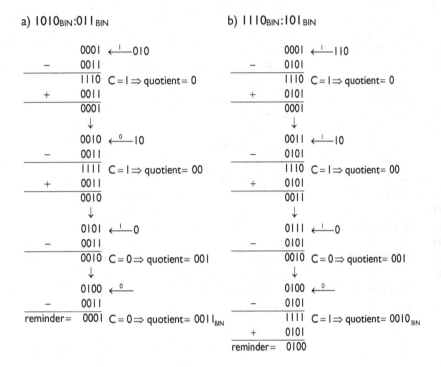

Implementation in code:

- input: B – higher byte of dividend, A – lower byte of dividend, R0 – divisor,
- output: R3 – higher byte of quotient, R2 – lower byte of quotient, B – higher byte of reminder, A – lower byte of reminder, OV – signal an attempt to divide by 0,
- exemplary value: $4487_{DEC}/100_{DEC}$.

```
 I      ;*********************************************************************
 2      ;* Division of BIN numbers 2 bytes/1 byte *
 3      ;* differential method II *
 4      ;*********************************************************************
 5      0011            n EQU 17
 6      0087            m EQU 135          ;17*256+135=4487 DEC
 7      0064            y EQU 100          ;y=100 DEC
 8
 9      0000: 75 F0 11  MOV B,#n           ;higher byte of dividend
10      0003: 74 87     MOV A,#m           ;lower byte of dividend
11      0005: 78 64     MOV R0,#y          ;divisor
```

```
12    0007: 12 00 0C      LCALL DIV_BIN16BY8DIFF
13                                                    ;result in {R3R2}-quotient,
14                                                    ; {BA}-reminder
15
16    000A: 80 FE         STOP: SJMP STOP
17    ;;------------------------------------------------------------------------------------
18    000C:               DIV_BIN16BY8DIFF:
19    000C: B8 00 03      CJNE R0,#0,DIVIDE
20    000F: D2 D2         SETB OV
21    0011: 22            RET
22    0012:               DIVIDE:
23    0012: 79 10         MOV R1,#16
24    0014: FE            MOV R6,A
25    0015:               LOOP:
26    0015: EE            MOV A,R6
27    0016: C3            CLR C
28    0017: 33            RLC A              ;<-lower byte of dividend
29    0018: C5 F0         XCH A,B
30    001A: 33            RLC A              ;<-higher byte of dividend
31    001B: C5 F0         XCH A,B
32    001D: FE            MOV R6,A
33    001E: CC            XCH A,R4
34    001F: 33            RLC A
35    0020: CD            XCH A,R5
36    0021: 33            RLC A
37    0022: CD            XCH A,R5           ;<-reminder<-C
38    0023: C3            CLR C
39    0024: 98            SUBB A,R0          ;reminder-dividend
40    0025: FC            MOV R4,A
41    0026: ED            MOV A,R5
42    0027: 94 00         SUBB A,#0
43    0029: FD            MOV R5,A
44    002A: 50 07         JNC NOT_LESS
45    002C: EC            MOV A,R4
46    002D: 28            ADD A,R0
47    002E: FC            MOV R4,A
48    002F: ED            MOV A,R5
49    0030: 34 00         ADDC A,#0
50    0032: FD            MOV R5,A
51    0033:               NOT_LESS:
52    0033: B3            CPL C
53    0034: CA            XCH A,R2
54    0035: 33            RLC A              ;<-quotient
55    0036: CB            XCH A,R3
56    0037: 33            RLC A
```

57	0038: CB	XCH A,R3
58	0039: CA	XCH A,R2
59	003A: D9 D9	DJNZ R1,LOOP
60	003C: ED	MOV A,R5
61	003D: F5 F0	MOV B,A
62	003F: EC	MOV A,R4
63	0040: 22	RET
64	;--- end of file ---	

Exercise 3.6: Perform division 10011_{BIN} by 1001_{BIN} using differential method II.

Exercise 3.7: Write a subroutine for dividing a 1-byte BIN number by a 1-byte number using differential method II.

Exercise 3.8*: Write a subroutine for dividing a 2-byte BIN number by a 2-byte number using:

  a. consecutive subtraction method,
  b. differential method II.

3.1.2 Working with Packed BCD

Addition of two P-BCD numbers is done similarly to adding BIN numbers, i.e. with the ADD instruction. In some cases it is necessary to correct the result, which is caused by application of radix 2 arithmetic to decimal numbers with radix 10! The next example shows the value of the carry bit between nibbles stored in the auxiliary carry AC flag of the processor, whose setting after addition is one of the conditions indicating the need to correct the result.

Example 3.13: Addition of two P-BCD numbers:

a.

$$84_{DEC}$$
$$+\ 13_{DEC}$$
$$\overline{97_{DEC}}$$

$$\overset{0}{10000100}_{P-BCD}$$
$$+\ 00010011_{P-BCD}$$
$$\overline{0\quad 10010111_{P-BCD}}$$

b.

$$25_{DEC}$$
$$+\ 37_{DEC}$$
$$\overline{62_{DEC}}$$

$$\overset{0}{00100101}_{P-BCD}$$
$$+\ 00110111_{P-BCD}$$
$$\overline{0\quad 01011100_{P-BCD}}$$
$$+\ 00000110_{P-BCD}$$
$$\overline{0\quad 01100010_{P-BCD}}$$

c.

$$94_{DEC}$$
$$+\ 16_{DEC}$$
$$\overline{110_{DEC}}$$

```
        0
   10010100_P-BCD
+  00010110_P-BCD
```
```
0   10101010
+   00000110
```
```
0   10110000
+   01100000
```
```
1   00010000_P-BCD
```

d.

$$87_{DEC}$$
$$+\ 19_{DEC}$$
$$\overline{106_{DEC}}$$

```
        1
   10000111_P-BCD
+  00011001_P-BCD
```
```
0   10100000
+   00000110
```
```
0   10100110
+   01100000
```
```
1   00000110_P-BCD
```

e.

$$85_{DEC}$$
$$+\ 92_{DEC}$$
$$\overline{177_{DEC}}$$

```
        0
   10000101_P-BCD
+  10010010_P-BCD
```
```
1   00010111
+   01100000
```
```
1   01110111_P-BCD
```

Adding two P-BCD numbers in the 8051 CPU is done by the ADD instruction followed by the DA A instruction, which automatically recognizes the necessity of result correction checking the following condition: if $A_{3...0} > 9$ or AC = 1, then $A \leftarrow A + 6$ after which, if $A_{7...4} > 9$ or C = 1, then $A \leftarrow A + 60h$. The flag C = 1 after correction indicates that the range for two-digit decimal number is exceeded, i.e. the result is greater than 99_{DEC}. However, if C is treated as a hundredth digit, then the result is correct, i.e. $>99_{DEC}$.

Implementation in code for case (b):

```
MOV A,#00100101B

ADD A,#00110111B

DA A                ;result in A
```

Exercise 3.9: Perform addition of P-BCD numbers:

a.

```
   10010110_P-BCD
+  00010101_P-BCD
?  ????????_P-BCD
```

b.

```
   10000100_P-BCD
+  01110011_P-BCD
?  ????????_P-BCD
```

Subtraction of two P-BCD numbers is performed by SUBB instruction after which decimal correction of the result must be considered. Unfortunately, in the 8051 CPU instruction list there is no equivalent of DA A that can be used after subtraction. So, the only solution is to substitute it with an appropriate piece of code. The correction works according to the following rule: if AC = 1, then $A \leftarrow A - 6$, followed by if C = 1, then

A←A – 60h. Setting the flag C = 1 after correction signals a negative (number below zero) result, which cannot be correctly interpreted in the sense of P-BCD. For this reason, avoid subtracting a larger number from a smaller number. Alternatively, the state of C = 1 may indicate the borrowing of a 100_{DEC} from the hundreds position, as presented in cases (d) and (e) of Example 3.14. In that case, the result is correct because a smaller number has been subtracted from a larger one. As in Example 31, the value of the AC flag is also signaled.

Example 3.14: Subtraction of two P-BCD numbers:

a.

$$48_{DEC}$$
$$-\ 16_{DEC}$$
$$32_{DEC}$$

```
              0
      01001000 P-BCD
    - 00010110 P-BCD
  0   00110010 P-BCD
```

b.

$$40_{DEC}$$
$$-\ 19_{DEC}$$
$$21_{DEC}$$

```
              1
      01000000 P-BCD
    - 00011001 P-BCD
      00100111
    - 00000110
  0   00100001 P-BCD
```

c.

$$72_{DEC}$$
$$-\ 35_{DEC}$$
$$37_{DEC}$$

```
              1
      01110010 P-BCD
    - 00110101 P-BCD
      00111101
    - 00000110
  0   00110111 P-BCD
```

d.

$$102_{DEC}$$
$$-\ 61_{DEC}$$
$$41_{DEC}$$

```
              0
      00000010 P-BCD
    - 01100001 P-BCD
  1   10100001
    - 01100000
  0   01000001 P-BCD
```

e.

$$107_{DEC}$$
$$-\ 89_{DEC}$$
$$18_{DEC}$$

```
              0
      00000111 P-BCD
    - 10001001 P-BCD
  1   01111110
    - 01100110
  0   00011000 P-BCD
```

The missing instruction 'decimal correction after subtraction' in 8051 CPU is emulated programmatically by the DA_A_S subroutine presented below. It starts with a label of this name and ends with a RET instruction. Since it will be reused also in other parts of the book, the occurring instruction 'LCALL DA_A_S' will always refer to the subprogram in the following listing, showing result of subtraction for data from the case d). Implementation in code:

- input: A – first number, B – second number,
- output: A – result, C – borrow from '100' position
- exemplary value: $00000010_{P\text{-}BCD} - 01100001_{P\text{-}BCD}$.

```
1     ;******************************************************************
2     ; * Subtraction of P-BCD numbers *
3     ;******************************************************************
4     0000: 74 02      MOV A,#02h          ;first number {C,A}=102 P-BCD
5     0002: 75 F0 61   MOV B,#61h          ;second number
6     0005: 12 00 0A   LCALL               SUB_PBCD
7                                          ;result in A
8     0008: 80 FE      STOP: SJMP STOP
9     ;--------------------------------------------------------------------
10    000A:            SUB_PBCD:
11    000A: C3         CLR C
12    000B: 95 F0      SUBB A,B
13    000D: 12 00 11   LCALL DA_A_S
14    0010: 22         RET
15    ;--------------------------------------------------------------------
16    0011:            DA_A_S:
17    ;emulation of 'Decimal Adjust after Subtraction'
18    0011: 85 D0 F0   MOV B,PSW
19    0014: 30 D6 03   JNB AC,SKIP
20    0017: C3         CLR C
21    0018: 94 06      SUBB A,#6
22    001A:            SKIP:
23    001A: 85 F0 D0   MOV PSW,B
24    001D: 50 03      JNC END
25    001F: C3         CLR C
26    0020: 94 60      SUBB A,#60h
27    0022:            END:
28    0022: 22         RET
29    ;--- end of file ---
```

Exercise 3.10: Perform subtraction of P-BCD numbers:

a.

$$10010010_{P-BCD}$$
$$- \ 10000111_{P-BCD}$$
$$? \quad ????????_{P-BCD}$$

b.

$$01100001_{P-BCD}$$
$$- \ 00100101_{P-BCD}$$
$$? \quad ????????_{P-BCD}$$

Multiplication and division of two P-BCD numbers are possible but complicated and generally produces abundant code. In practice, it is more convenient to convert P-BCD numbers into their BIN equivalents, perform the multiplication or division, and convert the result back into P-BCD form. The relevant algorithms are presented in chapters 2.2 and 3.1.1, and their final combination is left to the reader as a do-it-yourself task.

Exercise 3.11*: Write a subroutine for multiplication of two P-BCD numbers.

Exercise 3.12*: Write a subroutine for division of two P-BCD numbers.

3.1.3 Working with Unpacked BCD

Addition of two UP-BCD numbers is done similarly to P-BCD, adding the lower bytes with the ADD instruction and the higher bytes with the ADDC instruction, followed by the obligatory DA A 'decimal adjustment' instruction. Sometimes, a carry-over bit to the higher nibble of one or both bytes of the result may occur, resulting in an incorrect number in the UP-BCD sense. In such a case, the value F0h should be added to such a byte, considering the carry-over from the lower byte in operations on multi-byte numbers.

Example 3.15: Addition of two UP-BCD numbers:

a.

$$84_{DEC}$$
$$+ \ 13_{DEC}$$
$$\overline{97_{DEC}}$$

	0	0	
	00001000	00000100$_{UP-BCD}$	
+	00000001	00000011$_{UP-BCD}$	
0	00001001	00000111$_{UP-BCD}$	

b.

$$25_{DEC}$$
$$+ \ 37_{DEC}$$
$$\overline{62_{DEC}}$$

	0	0	
	00000010	00000101$_{UP-BCD}$	
+	00000011	00000111$_{UP-BCD}$	
0	00000101	00001100	
+	00000000	11110110	
0	00000110	00000010$_{UP-BCD}$	

c.

$$94_{DEC}$$
$$+ \ 16_{DEC}$$
$$\overline{110_{DEC}}$$

	0	0	
	00001001	00000100$_{UP-BCD}$	
+	00000001	00000110$_{UP-BCD}$	
0	00001010	00001010	
+	11110110	11110110	
1	00000001	00000000$_{UP-BCD}$	

d.

$$79_{DEC}$$
$$+ \ 98_{DEC}$$
$$\overline{177_{DEC}}$$

	1	1	
	00000111	00001001$_{UP-BCD}$	
+	00001001	00001000$_{UP-BCD}$	
0	00010000	00010001	
+	11110110	11110110	
1	00000111	00000111$_{UP-BCD}$	

Implementation in code:

- input: R1 – higher byte of first number, R0 – lower byte of first number, R3 – higher byte of second number, R2 – lower byte of second number,

- output: R1 – higher byte of result, R0 – lower byte of result,
- exemplary value: 00001001 00000100$_{\text{UP-BCD}}$ + 00000001 00000110$_{\text{UP-BCD}}$.

```
 I        ;*******************************************************************
 2        ;* Addition of UP-BCD numbers *
 3        ;*******************************************************************
 4        0000: 79 09           MOV R1,#09h
 5        0002: 78 04           MOV R0,#04h              ;first number
                                                        {R1,R0}=94
                                                        UP-BCD
 6        0004: 7B 01           MOV R3,#01h
 7        0006: 7A 06           MOV R2,#06h              ;second number
                                                        {R3,R2}=16
                                                        UP-BCD
 8        0008: 12 00 0D        LCALL ADD_UPBCD
 9                                                       ;result in {R1,R0}
10        000B: 80 FE           STOP: SJMP STOP
11        ;------------------------------------------------------------------
12        000D:                 ADD_UPBCD:
13        000D: E8              MOV A,R0
14        000E: 2A              ADD A,R2
15        000F: D4              DA A
16        0010: B4 09 02        CJNE A,#9,NOT_THE_SAME
17        0013: 80 05           SJMP SKIP
18        0015:                 NOT_THE_SAME:
19        0015: 40 03           JC SKIP
20        0017: 24 F0           ADD A,#0F0h
21        0019: 0B              INC R3
22        001A:                 SKIP:
23        001A: F8              MOV R0,A
24        001B: E9              MOV A,R1
25        001C: 2B              ADD A,R3
26        001D: D4              DA A
27        001E: B4 09 02        CJNE A,#9,SKIP1
28        0021: 80 04           SJMP END
29        0023:                 SKIP1:
30        0023: 40 02           JC END
31        0025: 24 F0           ADD A,#0F0h
32        0027:                 END:
33        0027: F9              MOV R1,A
34        0028: 22              RET
35        ;--- end of file ---
```

Exercise 3.13: Perform addition of two UP-BCD numbers:

a.

$$
\begin{array}{r}
00001001\ \ 00000100_{UP\text{-}BCD}\\
+\ \ 00000101\ \ 00000010_{UP\text{-}BCD}\\
\hline
?\ \ \ \ ????????\ \ \ \ ????????_{UP\text{-}BCD}
\end{array}
$$

b.

$$
\begin{array}{r}
00001001\ \ 00000111_{UP\text{-}BCD}\\
+\ \ 00000101\ \ 00001000_{UP\text{-}BCD}\\
\hline
?\ \ \ \ ????????\ \ \ \ ????????_{UP\text{-}BCD}
\end{array}
$$

Subtraction of two UP-BCD numbers requires a correction to be performed on the higher nibble of the result, just as after addition. Due to lack of a proper correction instruction after subtraction, the DA_A_S subroutine can be used as previously. Unfortunately, the correction algorithm, according to which the subroutine works, subtracts 60_{HEX}, among others, while it is required to subtract $F0_{HEX}$ for UP-BCD numbers. After executing the DA_A_S subroutine, the difference between $F0_{HEX}$ and 60_{HEX}, i.e. 90_{HEX}, must be subtracted additionally.

Example 3.16: Subtraction of two UP-BCD numbers:

a.

$$
\begin{array}{r}
48_{DEC}\\
-\ 16_{DEC}\\
\hline
32_{DEC}
\end{array}
$$

$$
\begin{array}{r}
0\qquad\qquad\quad 0\\
00000100\ \ 00001000_{UP\text{-}BCD}\\
-\ \ 00000001\ \ 00000110_{UP\text{-}BCD}\\
\hline
0\quad\ \ 00000011\ \ 00000010_{UP\text{-}BCD}
\end{array}
$$

b.

$$
\begin{array}{r}
72_{DEC}\\
-\ 35_{DEC}\\
\hline
37_{DEC}
\end{array}
$$

$$
\begin{array}{r}
0\qquad\qquad\quad 0\\
00000111\ \ 00000010_{UP\text{-}BCD}\\
-\ \ 00000011\ \ 00000101_{UP\text{-}BCD}\\
\hline
0\quad\ \ 00000011\ \ 11111101\\
-\ \ 00000000\ \ 11110110\\
\hline
0\quad\ \ 00000011\ \ 00000111_{UP\text{-}BCD}
\end{array}
$$

c.

$$
\begin{array}{r}
40_{DEC}\\
-\ 19_{DEC}\\
\hline
21_{DEC}
\end{array}
$$

$$
\begin{array}{r}
0\qquad\qquad\quad 1\\
00000100\ \ 00000000_{UP\text{-}BCD}\\
-\ \ 00000001\ \ 00001001_{UP\text{-}BCD}\\
\hline
0\quad\ \ 00000010\ \ 11110111\\
-\ \ 00000000\ \ 11110110\\
\hline
0\quad\ \ 00000010\ \ 00000001_{UP\text{-}BCD}
\end{array}
$$

d.

$$
\begin{array}{r}
102_{DEC}\\
-\ 61_{DEC}\\
\hline
41_{DEC}
\end{array}
$$

$$
\begin{array}{r}
0\qquad\qquad\quad 0\\
00000000\ \ 00000010_{UP\text{-}BCD}\\
-\ \ 00000110\ \ 00000001_{UP\text{-}BCD}\\
\hline
1\quad\ \ 11111010\ \ 00000001\\
-\ \ 11110110\ \ 00000000\\
\hline
0\quad\ \ 00000100\ \ 00000001_{UP\text{-}BCD}
\end{array}
$$

Implementation in code:

- input: R1 – higher byte of first number, R0 – lower byte of first number,
- R3 – higher byte of second number, R2 – lower byte of second number,
- output: R1 – higher byte of result, R0 – lower byte of result,
- exemplary value: $00000100\ 00001000_{UP\text{-}BCD} - 00000001$ $00000110_{UP\text{-}BCD}$.

```
 1      ;*****************************************************************
 2      ;* Subtraction of UP-BCD numbers *
 3      ;*****************************************************************
 4      0000: 79 04        MOV R1,#04h
 5      0002: 78 08        MOV R0,#08h              ;first number {R1,R0}
                                                       =48 UP-BCD
 6      0004: 7B 01        MOV R3,#01h
 7      0006: 7A 06        MOV R2,#06h              ;second number {R3,R2}
                                                       =16 UP-BCD
 8      0008: 12 00 0D     LCALL SUB_UPBCD
 9                                                  ;result in {R1,R0}
10      000B: 80 FE        STOP: SJMP STOP
11      ;-----------------------------------------------------------------
12      000D:              SUB_UPBCD:
13      000D: E8           MOV A,R0
14      000E: C3           CLR C
15      000F: 9A           SUBB A,R2
16      0010: C0 D0        PUSH PSW
17      0012: 12 00 31     LCALL DA_A_S
18      0015: B4 09 02     CJNE A,#9,NOT_THE_SAME
19      0018: 80 04        SJMP SKIP1
20      001A:              NOT_THE_SAME:
21      001A: 40 02        JC SKIP1
22      001C: 94 90        SUBB A,#90h
23      001E:              SKIP1:
24      001E: F8           MOV R0,A
25      001F: E9           MOV A,R1
26      0020: D0 D0        POP PSW
27      0022: 9B           SUBB A,R3
28      0023: 12 00 31     LCALL DA_A_S
29      0026: B4 09 02     CJNE A,#9,NOT_THE_SAME1
30      0029: 80 04        SJMP END1
31      002B:              NOT_THE_SAME1:
32      002B: 40 02        JC END1
33      002D: 94 90        SUBB A,#90h
```

```
34        002F:                    END1:
35        002F: F9                 MOV R1,A
36        0030: 22                 RET
37        ;----------------------------------------------------------------
38        0031:                    DA_A_S:
39                                 ;emulation of 'Decimal Adjust after Subtraction'
40        0031: 85 D0 F0           MOV B,PSW
41        0034: 30 D6 03           JNB AC,SKIP
42        0037: C3                 CLR C
43        0038: 94 06              SUBB A,#6
44        003A:                    SKIP:
45        003A: 85 F0 D0           MOV PSW,B
46        003D: 50 03              JNC END
47        003F: C3                 CLR C
48        0040: 94 60              SUBB A,#60h
49        0042:                    END:
50        0042: 22                 RET
51        ;--- end of file ---
```

Exercise 3.14: Perform addition of UP-BCD numbers:

a.

$$00001001 \quad 00000001_{UP-BCD}$$
$$- \ 00000110 \quad 00000011_{UP-BCD}$$
$$\overline{? \qquad ???????? \qquad ????????_{UP-BCD}}$$

b.

$$00000101 \quad 00000110_{UP-BCD}$$
$$- \ 00000100 \quad 00000010_{UP-BCD}$$
$$\overline{? \qquad ???????? \qquad ????????_{UP-BCD}}$$

Multiplication/division of two UP-BCD numbers is also troublesome. In practice, it is more convenient to convert UP-BCD numbers into their BIN equivalents, perform the multiplication/division and convert the result back into UP-BCD form. The corresponding algorithms are presented in subchapters 2.2 and 3.1.1.

Exercise 3.15*: Write a subroutine of multiplication of UP-BCD numbers.

Exercise 3.16*: Write a subroutine of division of UP-BCD numbers.

3.1.4 Working with Chars in ASCII

Addition of two numbers in ASCII code is performed similarly to UP-BCD format. In order not to introduce new rules for correcting the result, it is sufficient to clear the higher nibble of the adder before adding. The further procedure is the same as for UP-BCD.

Example 3.17: Addition of numbers in ASCII code:

a.

$$84_{DEC}$$
$$+ \ 13_{DEC}$$
$$97_{DEC}$$

00111000	00110100$_{ASCII}$
	↓
00001000	00000100
+ 00110001	0011001 1$_{ASCII}$
0 00111001	00110111$_{ASCII}$

b.

$$25_{DEC}$$
$$+ \ 37_{DEC}$$
$$62_{DEC}$$

	00110010	00110101$_{ASCII}$
		↓
	00000010	00000101
+	00110011	0011011 1$_{ASCII}$
0	00110101	00111100
+	00000000	11110110
0	00110110	00110010$_{ASCII}$

c.

$$94_{DEC}$$
$$+ \ 16_{DEC}$$
$$110_{DEC}$$

	00111001	00110100$_{ASCII}$
		↓
	00001001	00000100
+	00110001	00110110$_{ASCII}$
0	00111010	00111010
+	11110110	11110110
1	00110001	00110000$_{ASCII}$

d.

$$79_{DEC}$$
$$+ \ 98_{DEC}$$
$$177_{DEC}$$

	00110111	0011100 1$_{ASCII}$
		↓
	00000111	00001001
+	00111001	00111000$_{ASCII}$
0	01000000	01000001
+	11110110	11110110
1	00110111	0011011 1$_{ASCII}$

The following subroutine is a modified version of UP-BCD addition, with the conversion from ASCII to UP-BCD is made easy by using an extra ANL A,#0Fh instruction.

Implementation in code:

- input: R1 – higher byte of first number, R0 – lower byte of first number, R3 – higher byte of second number, R2 – lower byte of second number,
- output: R1 – higher byte of result, R0 – lower byte of result,
- exemplary value: 00111001 00110100$_{ASCII}$ + 00110001 00110110$_{ASCII}$.

```
1        ;**********************************************************************
         ;
2        ;* Addition of ASCII numbers *
3        ;**********************************************************************
         ;
4        0000: 79 39        MOV R1,#39h
```

5	0002: 78 34	MOV R0,#34h	;first number {R1,R0}=94 ASCII
6	0004: 7B 31	MOV R3,#31h	
7	0006: 7A 36	MOV R2,#36h	;second number {R3,R2}=16 ASCII
8	0008: 12 00 0D	LCALL ASCII_ADD	
9			;result in {R1,R0}
10	000B: 80 FE	STOP: SJMP STOP	
11	;---
12	000D:	ASCII_ADD:	
13	000D: E9	MOV A,R1	
14	000E: 54 0F	ANL A,#0Fh	
15	0010: F9	MOV R1,A	
16	0011: E8	MOV A,R0	
17	0012: 54 0F	ANL A,#0Fh	
18	0014: 2A	ADD A,R2	
19	0015: D4	DA A	
20	0016: B4 09 02	CJNE A,#9,NOT_THE_SAME	
21	0019: 80 05	SJMP SKIP	
22	001B:	NOT_THE_SAME:	
23	001B: 40 03	JC SKIP	
24	001D: 24 F0	ADD A,#0F0h	
25	001F: 0B	INC R3	
26	0020:	SKIP:	
27	0020: F8	MOV R0,A	
28	0021: E9	MOV A,R1	
29	0022: 2B	ADD A,R3	
30	0023: D4	DA A	
31	0024: B4 09 02	CJNE A,#9,NOT_THE_SAME1	
32	0027: 80 04	SJMP END	
33	0029:	NOT_THE_SAME1:	
34	0029: 40 02	JC END	
35	002B: 24 F0	ADD A,#0F0h	
36	002D:	END:	
37	002D: F9	MOV R1,A	
38	002E: 22	RET	
39	;--- end of file ---		

Exercise 3.17: Perform addition of numbers in ASCII code:

a.

$$00110011 \ 00110010_{ASCII}$$
$$+ \ 00110111 \ 00110011_{ASCII}$$

| ? | ???????? | ????????$_{ASCII}$ |

b.

$$00110101 \ 00110110_{ASCII}$$
$$+ \ 00110111 \ 00110011_{ASCII}$$

| ? | ???????? | ????????$_{ASCII}$ |

Subtraction of two numbers in ASCII code is performed similarly to UP-BCD. To avoid new rules for correcting the result, just clear the higher nibble of the second number before the subtraction. The further procedure is the same as for UP-BCD.

Example 3.18: Subtraction of two numbers in ASCII code:

a.

$$48_{DEC}$$
$$-\ 16_{DEC}$$
$$\overline{32_{DEC}}$$

```
  00110100  00111000_ASCII
- 00110001  00110110_ASCII
              ↓
⁸0 00110100  00111000_ASCII
-  00000001  00000110
 0 00110011  00110010_ASCII
```

b.

$$72_{DEC}$$
$$-\ 35_{DEC}$$
$$\overline{37_{DEC}}$$

```
  00110111  00110010_ASCII
- 00110011  00110101_ASCII
              ↓
0 00110111  00110010_ASCII
- 00000011  00000101
0 00110100  00101101
- 00000000  11110110
0 00110011  00110111_ASCII
```

c.

$$40_{DEC}$$
$$-\ 19_{DEC}$$
$$\overline{21_{DEC}}$$

```
  00110100  00110000_ASCII
  00110001  00111001_ASCII
              ↓
  00110100  00110000_ASCII
- 00000001  00001001
0 00110011  00100111
- 00000000  11110110
0 00110010  00110001_ASCII
```

d.

$$102_{DEC}$$
$$-\ 61_{DEC}$$
$$\overline{41_{DEC}}$$

```
  00110000  00110010_ASCII
- 00110110  00110001_ASCII
              ↓
  00110000  00110010_ASCII
- 00000110  00000001
1 00101010  00110001
- 11110110  00000000
0 00110100  00110001_ASCII
```

Implementation in code:

- input: R1 – higher byte of first number, R0 – lower byte of first number,
- R3 – higher byte of second number, R2 – lower byte of second number,
- output: R1 – higher byte of result, R0 – lower byte of result,
- exemplary value: 00110100 00110000$_{ASCII}$ + 00110001 00111001$_{ASCII}$.

```
1    ;*********************************************************************
2    ;* Subtraction of ASCII numbers *
3    ;*********************************************************************
4    0000: 79 34         MOV R1,#34h
5    0002: 78 30         MOV R0,#30h                    first number
                                                        {R1,R0}=40 ASCII
6    0004: 7B 31         MOV R3,#31h
7    0006: 7A 39         MOV R2,#39h                    ;second number
                                                        {R3,R2}=19 ASCII
8    0008: 12 00 0D      LCALL ASCII_SUB
9                                                       ;result in {R1,R0}
10   000B: 80 FE         STOP: SJMP STOP
11   ;--------------------------------------------------------------------------------
12   000D:               ASCII_SUB:
13   000D: EB            MOV A,R3
14   000E: 54 0F         ANL A,#0Fh
15   0010: FB            MOV R3,A
16   0011: EA            MOV A,R2
17   0012: 54 0F         ANL A,#0Fh
18   0014: FA            MOV R2,A
19   0015: E8            MOV A,R0
20   0016: C3            CLR C
21   0017: 9A            SUBB A,R2
22   0018: 12 00 37      LCALL DA_A_S
23   001B: B4 30 02      CJNE A,#30h,NOT_THE_SAME
24   001E: 80 05         SJMP SKIP1
25   0020:               NOT_THE_SAME:
26   0020: 50 03         JNC SKIP1
27   0022: C3            CLR C
28   0023: 94 F0         SUBB A,#0F0h
29   0025:               SKIP1:
30   0025: F8            MOV R0,A
31   0026: E9            MOV A,R1
32   0027: 9B            SUBB A,R3
33   0028: 12 00 37      LCALL DA_A_S
34   002B: B4 30 02      CJNE A,#30h,NOT_THE_SAME1
35   002E: 80 05         SJMP END1
36   0030:               NOT_THE_SAME1:
37   0030: 50 03         JNC END1
38   0032: C3            CLR C
39   0033: 94 F0         SUBB A,#0F0h
40   0035:               END1:
41   0035: F9            MOV R1,A
42   0036: 22            RET
43
```

```
44      ;-----------------------------------------------------------------------------------------------
45      0037:                   DA_A_S:
46                              ;emulation of 'Decimal Adjust after Subtraction'
47      0037: 85 D0 F0          MOV B,PSW
48      003A: 30 D6 03          JNB AC,SKIP
49      003D: C3               CLR C
50      003E: 94 06            SUBB A,#6
51      0040:                   SKIP:
52      0040: 85 F0 D0          MOV PSW,B
53      0043: 50 03            JNC END
54      0045: C3               CLR C
55      0046: 94 60            SUBB A,#60h
56      0048:                   END:
57      0048: 22               RET
58      ;--- end of file ---
```

Exercise 3.18: Perform subtraction of numbers in ASCII code:

a.

00111001	00110010$_{ASCII}$
− 00110111	00110001$_{ASCII}$
? ????????	????????$_{ASCII}$

b.

00110101	00110110$_{ASCII}$
− 00110111	00110010$_{ASCII}$
? ????????	????????$_{ASCII}$

Multiplying and dividing two numbers in ASCII code is complicated. In practice, it is more convenient to convert ASCII numbers into their BIN equivalents, perform the multiplication or division, and convert the result back into ASCII code. The corresponding algorithms are presented in chapters 2.2 and 3.1.1.

Exercise 3.19*: Write a subroutine of multiplication of numbers in ASCII code.

Exercise 3.20*: Write a subroutine of division of numbers in ASCII code.

3.2 OPERATIONS ON SIGNED NUMBERS

3.2.1 Working with Sign-magnitude

Addition of two SM numbers requires consideration of the sign of both numbers. One of two cases may occur:

Table 3.1 Rules for Adding Numbers in the Sign-magnitude

Sign of A = sign of B		$\|S\|=\|A\|+\|B\|$ sign of S = sign of A
sign of A ≠ sign of B	If $\|A\| \geq \|B\|$ then: $\|S\|=\|A\| - \|B\|$ sign of S = sign of A	if $\|A\|<\|B\|$ then: $\|S\|=\|B\| - \|A\|$ sign of S = /sign of A

- if the signs of the two numbers match, then the modulus of the result is the sum of the moduli of the numbers and the sign of the result is equal to their sign;
- if the signs of both numbers differ, then the module of the result is determined by subtracting from the module of the larger number the module of the smaller number and the sign of the result is equal to the sign of the larger number.

These rules are simply illustrated in Table 3.1, where A is the first number, B the second number, S is the sum, i.e. S = A + B and '/' mark means the sign inversion.

Example 3.19: Addition of two SM numbers:

a.

$$
\begin{array}{cc}
(-3)_{DEC} & \underline{1}001\,1_{SM} \\
+\ (-5)_{DEC} & +\ \underline{1}010\,1_{SM} \\
\hline
-8_{DEC} & \underline{1}\ 1000_{SM}
\end{array}
$$

b.

$$
\begin{array}{cc}
(+3)_{DEC} & \underline{0}001\,1_{SM} \\
+\ (+5)_{DEC} & +\ \underline{0}010\,1_{SM} \\
\hline
+8_{DEC} & \underline{0}1000_{SM}
\end{array}
$$

c.

$$
\begin{array}{cc}
(-3)_{DEC} & \underline{1}001\,1_{SM} \\
+\ (+5)_{DEC} & +\ \underline{0}010\,1_{SM} \\
\hline
+2_{DEC} & ?
\end{array}
$$

$\xrightarrow{|A|<|B|}$

$$
\begin{array}{l}
0101 \\
-\ 0011 \\
\hline
\underline{0}\ \ 0010_{SM}
\end{array}
$$

d.

$$
\begin{array}{cc}
(-5)_{DEC} & \underline{1}010\,1_{SM} \\
+\ (+3)_{DEC} & +\ \underline{0}001\,1_{SM} \\
\hline
-2_{DEC} & ?
\end{array}
$$

$\xrightarrow{|A|>|B|}$

$$
\begin{array}{l}
0101 \\
-\ 0011 \\
\hline
\underline{1}\ \ 0010_{SM}
\end{array}
$$

Implementation in code:

- input: A – first number, B – second number,
- output: A – result, OV – result out of the range,
- exemplary value: $10001100B_{SM} + 10010011B_{SM}$.

```
 1    ;********************************************************************
 2    ;* Addition of SM numbers *
 3    ;********************************************************************
 4    008C                    n EQU 10001100B          ;-12 SM
 5    0093                    m EQU 10010011B          ;-19 SM
 6
 7    0000: 74 8C             MOV A,#n                 ;first number
 8    0002: 75 F0 93          MOV B,#m                 ;second number
 9    0005: 12 00 0A          LCALL SM_ADD
10                                                     ;result in A
11    0008: 80 FE             STOP: SJMP STOP
12    ;---------------------------------------------------------------------------------------
13    000A:                   SM_ADD:
14    000A: A2 E7             MOV C,ACC.7
15    000C: 92 D5             MOV PSW.5,C              ;sign of n
16    000E: C0 E0             PUSH ACC
17    0010: 65 F0             XRL A,B
18    0012: A2 E7             MOV C,ACC.7              ;sign of m
19    0014: 53 F0 7F          ANL B,#01111111B         ;|m|
20    0017: D0 E0             POP ACC
21    0019: 54 7F             ANL A,#01111111B         ;|n|
22    001B: 50 18             JNC SIGN_THE_SAME
23    001D: B5 F0 02          CJNE A,B,SKIP
24    0020: 80 02             SJMP NOT_LESS
25    0022:                   SKIP:
26    0022: 40 07             JC LESS
27    0024:                   NOT_LESS:
28    0024: C3                CLR C
29    0025: 95 F0             SUBB A,B
30    0027: A2 D5             MOV C,PSW.5
31    0029: 80 12             SJMP END
32    002B:                   LESS:
33    002B: C3                CLR C
34    002C: C5 F0             XCH A,B
35    002E: 95 F0             SUBB A,B
36    0030: A2 D5             MOV C,PSW.5
37    0032: B3                CPL C
38    0033: 80 08             SJMP END
39    0035:                   SIGN_THE_SAME:
40    0035: 25 F0             ADD A,B
41    0037: A2 E7             MOV C,ACC.7
42    0039: 92 D2             MOV OV,C
43    003B: A2 D5             MOV C,PSW.5
```

44	003D:	END:
45	003D: 92 E7	MOV ACC.7,C
46	003F: 22	RET
47	;--- end of file ---	

Carry-over to the sign bit of the result (here the highest bit of the accumulator) of adding numbers with the same signs means exceeding the range for single-byte numbers in SM. This case is indicated by setting the OV flag. The result in accumulator should be discarded then!

Exercise 3.21: Perform addition of SM numbers:

a.
$$\underline{1}\,100_{SM}$$
$$+\ \underline{1}\,111_{SM}$$
$$????_{SM}$$

b.
$$\underline{0}\,100_{SM}$$
$$+\ \underline{0}\,111_{SM}$$
$$????_{SM}$$

c.
$$\underline{1}\,100_{SM}$$
$$+\ \underline{0}\,111_{SM}$$
$$????_{SM}$$

d.
$$\underline{0}\,100_{SM}$$
$$+\ \underline{1}\,111_{SM}$$
$$????_{SM}$$

The case of different signs can be solved in another way by using complements. The negative number is replaced by its 1's or 2's complement. In next step the addition is performed, and the result is corrected according to the rules for using complements, described in chapter 3.1 – look for 'Subtraction of two BIN numbers by means of complements'.

Example 3.20: Addition of two SM numbers for case of different signs by using complements:

a.
$$(- 3)_{DEC} \quad \underline{1}\,001\,1_{SM}$$
$$+\ (+ 5)_{DEC} +\ \underline{0}\,010\,1_{SM}$$
$$\overline{+2_{DEC} \qquad\qquad ?}$$

$$1100_{1's}$$
$$+\ 0101$$
$$\overline{\mathbf{1}\ 0001}$$
$$+\ \ 0001$$
$$\overline{\underline{0}\ 0010_{SM}}$$

or

$$1101_{U2}$$
$$+\ 0101$$
$$\overline{\mathbf{1}\ 0010}$$
$$\Downarrow$$
$$\underline{0}\ 0010_{SM}$$

b.
$$(+ 3)_{DEC} \quad \underline{0}\,001\,1_{SM}$$
$$+\ (- 5)_{DEC} +\ \underline{1}\,010\,1_{SM}$$
$$\overline{-2_{DEC} \qquad\qquad ?}$$

$$0011$$
$$+\ 1010_{U1}$$
$$\overline{\cancel{0}\ 1101}$$
$$\Downarrow$$
$$\underline{1}\ 0010_{SM}$$

or

$$0011$$
$$+\ 1011_{2's}$$
$$\overline{\cancel{0}\ 1110}$$
$$\Downarrow$$
$$0001$$
$$+\ 0001$$
$$\overline{\underline{1}\ 0010_{SM}}$$

Implementation in code using 1's complement:

- input: A – first number, B – second number,
- output: A – result, OV – result out of the range,
- exemplary value: $10001100_{SM} + 10010011_{SM}$.

```
 1    ;*********************************************************************
 2    ;* Addition of SM numbers by I's complement *
 3    ;*********************************************************************
 4    008C                    n EQU 10001100B              ;-12 SM
 5    0093                    m EQU 10010011B              ;-19 SM
 6
 7    0000: 74 8C             MOV A,#n                     ;first number
 8    0002: 75 F0 93          MOV B,#m                     ;second number
 9    0005: 12 00 0A          LCALL SM_ADD_BYIS
10                                                         ;result in A
11    0008: 80 FE             STOP: SJMP STOP
12    ;-----------------------------------------------------------------------------
13    000A:                   SM_ADD_BYIS:
14    000A: A2 E7             MOV C,ACC.7
15    000C: 92 D5             MOV PSW.5,C                  ;sign of n
16    000E: C0 E0             PUSH ACC
17    0010: 65 F0             XRL A,B
18    0012: A2 E7             MOV C,ACC.7
19    0014: D0 E0             POP ACC
20    0016: 50 20             JNC SIGN_THE_SAME
21    0018:                   NOT_THE_SAME:
22    0018: C5 F0             XCH A,B
23    001A: 30 E7 03          JNB ACC.7,POSITIVE_M
24    001D: F4                CPL A
25    001E: B2 D1             CPL PSW.1
26    0020:                   POSITIVE_M:
27    0020: C5 F0             XCH A,B
28    0022: 30 E7 03          JNB ACC.7,POSITIVE_N
29    0025: F4                CPL A
30    0026: B2 D1             CPL PSW.1
31    0028:                   POSITIVE_N:
32    0028: 30 D1 0D          JNB PSW.1,SIGN_THE_SAME
33    002B: 25 F0             ADD A,B
34    002D: 30 E7 05          JNB ACC.7,SKIPI
35    0030: 04                INC A
36    0031: B2 E7             CPL ACC.7
```

37	0033: 80 0D	SJMP END
38	0035:	SKIP1:
39	0035: F4	CPL A
40	0036: 80 0A	SJMP END
41	0038:	SIGN_THE_SAME:
42	0038: 25 F0	ADD A,B
43	003A: A2 E7	MOV C,ACC.7
44	003C: 92 D2	MOV OV,C
45	003E: A2 D5	MOV C,PSW.5
46	0040: 92 E7	MOV ACC.7,C
47	0042:	END:
48	0042: 22	RET
49	;--- end of file ---	

Below code presents the implementation of addition by using 2's complement.
Implementation in code using 2's complement:

- input: A – first number, B – second number,
- output: A – result, OV – result out of the range,
- exemplary value: $10001100_{SM} + 10010011_{SM}$.

```
 1        ;**************************************************************
 2        ;* Addition of SM numbers by 2's complement *
 3        ;**************************************************************
 4   008C            n EQU 10001100B              ;-12 SM
 5   0093            m EQU 10010011B              ;-19 SM
 6
 7   0000: 74 8C     MOV A,#n                     ;first number
 8   0002: 75 F0 93  MOV B,#m                     ;second number
 9   0005: 12 00 0A  LCALL SM_ADD_BY2S
10                                                ;result in A
11   0008: 80 FE     STOP: SJMP STOP
12        ;-----------------------------------------------------------------------
13   000A:           SM_ADD_BY2S:
14   000A: A2 E7     MOV C,ACC.7
15   000C: 92 D5     MOV PSW.5,C                  ;sign of n
16   000E: C0 E0     PUSH ACC
17   0010: 65 F0     XRL A,B
18   0012: A2 E7     MOV C,ACC.7
19   0014: D0 E0     POP ACC
20   0016: 50 22     JNC SIGN_THE_SAME
```

```
21    0018:                   NOT_THE_SAME:
22    0018: C5 F0             XCH A,B
23    001A: 30 E7 04          JNB ACC.7,POSITIVE_M
24    001D: F4                CPL A
25    001E: 04                INC A
26    001F: B2 D1             CPL PSW.1
27    0021:                   POSITIVE_M:
28    0021: C5 F0             XCH A,B
29    0023: 30 E7 04          JNB ACC.7,POSITIVE_N
30    0026: F4                CPL A
31    h0027: 04               INC A
32    0028: B2 D1             CPL PSW.1
33    002A:                   POSITIVE_N:
34    002A: 30 D1 0D          JNB PSW.1,SIGN_THE_SAME
35    002D: 25 F0             ADD A,B
36    002F: 30 E7 04          JNB ACC.7,SKIP1
37    0032: B2 E7             CPL ACC.7
38    0034: 80 0E             SJMP END
39    0036:                   SKIP1:
40    0036: F4                CPL A
41    0037: 04                INC A
42    0038: 80 0A             SJMP END
43    003A:                   SIGN_THE_SAME:
44    003A: 25 F0             ADD A,B
45    003C: A2 E7             MOV C,ACC.7
46    003E: 92 D2             MOV OV,C
47    0040: A2 D5             MOV C,PSW.5
48    0042: 92 E7             MOV ACC.7,C
49    0044:                   END:
50    0044: 22                RET
51    ;--- end of file ---
```

Subtraction of two SM numbers requires consideration of the sign of both numbers. One of two cases may occur:

- if the signs of the two numbers differ, then the modulus of the result is the sum of the moduli of the numbers and the sign of the result is equal to their sign;
- if the signs of both numbers match, then the module of the result is determined by subtracting from the module of the larger number the module of the smaller number and the sign of the result is equal to the sign of the larger number.

These rules are simply illustrated in Table 3.2, where A is the first number, B the second number, D is the difference, i.e. D = A – B and '/' mark means the sign inversion.

Table 3.2 Rules for Subtracting Numbers in the Sign-magnitude

sign of A ≠ sign of B		$R=\|A\|+\|B\|$ sign of D = sign of A
sign of A = sign of B	if $\|A\| \geq \|B\|$ then: $\|D\|=\|A\| - \|B\|$ sign of D = sign of A	if $\|A\| < \|B\|$ then: $\|R\|=\|B\| - \|A\|$ sign R = /sign of A

Example 3.21: Subtraction of two SM numbers:

a.

$$\begin{array}{ll} (+3)_{DEC} & \underline{0}001 1_{SM} \\ - (-5)_{DEC} & - \underline{1}0101_{SM} \\ \hline +8_{DEC} & \underline{0}1000_{SM} \end{array}$$

b.

$$\begin{array}{ll} (-3)_{DEC} & \underline{1}001 1_{SM} \\ - (+5)_{DEC} & + \underline{0}0101_{SM} \\ \hline -8_{DEC} & \underline{1}1000_{SM} \end{array}$$

c.

$$\begin{array}{ll} (+3)_{DEC} & \underline{0}001 1_{SM} \\ - (+5)_{DEC} & - \underline{0}0101_{SM} \\ \hline -2_{DEC} & ? \end{array} \xrightarrow{|A|<|B|} \begin{array}{l} 0101 \\ - 0011 \\ \hline \underline{1} \ 0010_{SM} \end{array}$$

d.

$$\begin{array}{ll} (-5)_{DEC} & \underline{1}0101_{SM} \\ - (-3)_{DEC} & - \underline{1}001 1_{SM} \\ \hline -2_{DEC} & ? \end{array} \xrightarrow{|A|>|B|} \begin{array}{l} 0101 \\ - 0011 \\ \hline \underline{1} \ 0010_{SM} \end{array}$$

Implementation in code:

- input: A – first number, B – second number,
- output: A – result, OV – result out of the range,
- exemplary value: $10001100_{SM} - 10010011_{SM}$.

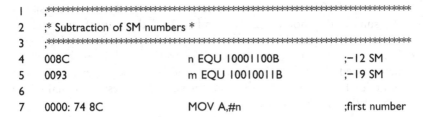

```
 1    ;*********************************************************************
 2    ;* Subtraction of SM numbers *
 3    ;*********************************************************************
 4    008C            n EQU 10001100B              ;-12 SM
 5    0093            m EQU 10010011B              ;-19 SM
 6
 7    0000: 74 8C     MOV A,#n                     ;first number
```

```
8     0002: 75 F0 93        MOV B,#m                  ;second number
9     0005: 12 00 0A        LCALL SM_SUB
10                                                    ;result in A
11    0008: 80 FE           STOP: SJMP STOP
12    ;-----------------------------------------------------------------------------------------
13    000A:                 SM_SUB:
14    000A: A2 E7           MOV C,ACC.7
15    000C: 92 D5           MOV PSW.5,C                ;sign of n
16    000E: C0 E0           PUSH ACC
17    0010: 65 F0           XRL A,B
18    0012: A2 E7           MOV C,ACC.7                ;sign of m
19    0014: 53 F0 7F        ANL B,#01111111B          ;|m|
20    0017: D0 E0           POP ACC
21    0019: 54 7F           ANL A,#01111111B          ;|n|
22    001B: 40 18           JC NOT_THE_SAME
23    001D:                 SIGN_THE_SAME:
24    001D: B5 F0 02        CJNE A,B,SKIP
25    0020: 80 02           SJMP NOT_LESS
26    0022:                 SKIP:
27    0022: 40 07           JC LESS
28    0024:                 NOT_LESS:
29    0024: C3              CLR C
30    0025: 95 F0           SUBB A,B
31    0027: A2 D5           MOV C,PSW.5
32    0029: 80 12           SJMP END
33    002B:                 LESS:
34    002B: C3              CLR C
35    002C: C5 F0           XCH A,B
36    002E: 95 F0           SUBB A,B
37    0030: A2 D5           MOV C,PSW.5
38    0032: B3              CPL C
39    0033: 80 08           SJMP END
40    0035:                 NOT_THE_SAME:
41    0035: 25 F0           ADD A,B
42    0037: A2 E7           MOV C,ACC.7
43    0039: 92 D2           MOV OV,C
44    003B: A2 D5           MOV C,PSW.5
45    003D:                 END:
46    003D: 92 E7           MOV ACC.7,C
47    003F: 22              RET
48    ;--- end of file ---
```

This time, the carry-over to the sign bit of the result (the highest bit of the accumulator) of subtraction of two numbers with different signs means exceeding the range for single-byte numbers in SM. This case is indicated by setting the OV flag. The result in the accumulator must be discarded now!

Exercise 3.22: Perform subtraction of SM numbers:

a.

$$\underline{1}\,100_{SM}$$
$$-\ \underline{1}\,111_{SM}$$
$$\overline{????_{SM}}$$

b.

$$\underline{0}\,100_{SM}$$
$$-\ \underline{0}\,111_{SM}$$
$$\overline{????_{SM}}$$

c.

$$\underline{1}\,100_{SM}$$
$$-\ \underline{0}\,111_{SM}$$
$$\overline{????_{SM}}$$

d.

$$\underline{0}\,100_{SM}$$
$$-\ \underline{1}\,111_{SM}$$
$$\overline{????_{SM}}$$

Multiplication of two SM numbers is performed similarly as to BIN numbers. The number modules are multiplied, the result bit is calculated as an XOR function of the number sign bits. The result of single-byte number multiplication fits into two bytes.

Implementation in code:

- input: A – multiplicand, B – multiplier,
- output: B – higher byte of result, A – lower byte of result,
- exemplary value: $10001100_{SM} * 10010011_{SM}$.

```
 1     ;**************************************************************
 2     ;* Multiplication of SM numbers *
 3     ;**************************************************************
 4     008C              n EQU 10001100B          ;-12 SM
 5     0093              m EQU 10010011B          ;-19 SM
 6
 7     0000: 74 8C       MOV A,#n                 ;multiplicand
 8     0002: 75 F0 93    MOV B,#m                 ;multiplier
 9     0005: 12 00 0A    LCALL SM_MUL
10                                                ;result in {B,A}
11     0008: 80 FE       STOP: SJMP STOP
12     ;--------------------------------------------------------------
13     000A:             SM_MUL:
14     000A: C0 E0       PUSH ACC
15     000C: 65 F0       XRL A,B
16     000E: A2 E7       MOV C,ACC.7
```

```
17    0010: 92 D5              MOV PSW.5,C
18    0012: 53 F0 7F           ANL B,#01111111B
19    0015: D0 E0              POP ACC
20    0017: 54 7F              ANL A,#01111111B          ;|n|
21    0019: A4                 MUL AB
22    001A: A2 D5              MOV C,PSW.5
23    001C: 92 F7              MOV B.7,C                 ;sign of result
24    001E: 22                 RET
25    ;--- end of file ---
```

The division of two SM numbers is done similarly to the division of BIN numbers. The modules of number are divided and the sign bit of result is calculated as an XOR function of the sign bits of both numbers. The result of dividing single-byte numbers consists of a quotient byte and a remainder byte.

Implementation in code:

- input number: A – dividend, B – divisor,
- output number: A – quotient, B – reminder, OV – division by zero,
- exemplary value: $00010100_{SM}/10010000_{SM}$.

```
 1    ;***************************************************************
 2    ;* Division of SM *
 3    ;***************************************************************
 4    0014                     n EQU 00010100B           ;+20 SM
 5    0090                     m EQU 10010000B           ;-16 SM
 6
 7    0000: 74 14              MOV A,#n                  ;dividend
 8    0002: 75 F0 90           MOV B,#m                  ;divisor
 9    0005: 12 00 0A           LCALL                     SM_DIV
10                                                       ;result in A-quotient
11                                                       ;B-reminder
12    0008: 80 FE              STOP: SJMP STOP
13    ;-------------------------------------------------------------------
14    000A:                    SM_DIV:
15    000A: C0 E0              PUSH ACC
16    000C: 65 F0              XRL A,B
17    000E: A2 E7              MOV C,ACC.7
18    0010: 92 D5              MOV PSW.5,C
19    0012: 53 F0 7F           ANL B,#01111111B          ;|m|
20    0015: D0 E0              POP ACC
21    0017: 54 7F              ANL A,#01111111B          ;|n|
22    0019: 84                 DIV AB
```

```
23      001A: A2 D5          MOV C,PSW.5
24      001C: 92 E7          MOV ACC.7,C                ;sign of result
25      001E: 22             RET
26      ;--- end of file ---
```

An attempt to divide by 0 is signaled by setting the OV flag. The contents of registers A and B are then meaningless.

3.2.2 Working with 2's Complement

Compared to other systems for representing signed numbers 2's complement has the advantage that the basic arithmetic operations of addition, subtraction and multiplication are identical to those for unsigned binary numbers. It is true as long as the inputs are represented in the same number of bits as the output, and result can be expressed properly in this number of bits. This property makes the hardware or software implementation simpler, especially for higher-precision arithmetic.

Addition of two 2's complement numbers is done similarly as for BINs by adding the sign bits as well. The carry-over bit must be obligatorily discarded. If the range is exceeded for numbers in 2's, the processor's OV flag is set.

In the next example, the carry bit was crossed out and the carry value from the two highest bits was distinguished, based on which the processor determines the state of the OV flag.

Example 3.22: Addition of two 2's numbers:

a.

$$
\begin{array}{r}
(+7)_{DEC} \\
+ \quad (-4)_{DEC} \\
\hline
+3_{DEC}
\end{array}
$$

$$
\begin{array}{r}
11 \\
0111_{2's} \\
+ \quad 1100_{2's} \\
\hline
1 \; 0011_{2's}
\end{array}
$$

b.

$$
\begin{array}{r}
(+3)_{DEC} \\
+ \quad (+6)_{DEC} \\
\hline
+9_{DEC}
\end{array}
$$

$$
\begin{array}{r}
01 \\
0011_{2's} \\
+ \quad 0110_{2's} \\
\hline
0 \; 1001_{2's}
\end{array}
$$

In case (b), a result was out of the range assuming 4-bit 2's numbers. To get the correct result, repeat the operation for 5-bit numbers as below:

b.*

$$
\begin{array}{r}
00 \\
00011_{2's} \\
+ \quad 00110_{2's} \\
\hline
0 \; 01001_{2's}
\end{array}
$$

Exercise 3.23: Perform addition of 2's numbers:

a.   b.

$\quad 1001_{2's}$ $\quad 1011_{2's}$

$+\ 1111_{2's}$ $+\ 0110_{2's}$

$\quad ????_{2's}$ $\quad ????_{2's}$

Subtraction of two 2's complement numbers is done similarly as for BINs by subtracting the sign bits as well. The borrow bit must be obligatorily discarded. If the range is exceeded for numbers in 2's, the processor's OV flag is set.

In the next example, the borrow bit was crossed out and the borrowed value from the two highest bits were distinguished, based on which the processor determines the state of the OV flag.

Example 3.23: Subtraction of two 2's numbers:

a.   b.

$\quad (+6)_{DEC}$ $\quad (-2)_{DEC}$

$-\ (-2)_{DEC}$ $-\ (+5)_{DEC}$

$\quad +8_{DEC}$ $\quad -7_{DEC}$

$\ \ {}^{10}$ $\ \ {}^{00}$

$\quad 0110_{2's}$ $\quad 1110_{2's}$

$-\ 1110_{2's}$ $-\ 0101_{2's}$

$\overline{1\ 1000_{2's}}$ $\overline{\not{0}\ 1001_{2's}}$

In case (a), a result was out of the range assuming 4-bit 2's numbers. To get the correct result repeat the operation for 5-bits numbers as below:

a.*

$\ \ {}^{11}$

$\quad 00110_{2's}$

$-\ 11110_{2's}$

$\overline{\not{0}\ 01000_{2's}}$

Exercise 3.24: Perform subtraction of 2's numbers:

a.   b.

$\quad 1101_{2's}$ $\quad 0101_{2's}$

$-\ 0011_{2's}$ $-\ 1100_{2's}$

$\quad ????_{2's}$ $\quad ????_{2's}$

 **REMEMBER!**

If after the addition operation the carry-over or after the subtraction operation, the borrowings from the two highest bits are different then the overflow occurred. It means that the result cannot be expressed by the provided number of bits. The result must be discarded or the operation on numbers with more bits (at least one extra bit is needed) must be performed again. The overflow is signaled by the processor by setting its OV flag.

Multiplication of two 2's numbers – sign change method – the method uses the principle of converting every negative number to a positive one (see chapter 2.4). Then they are multiplied as positive numbers without the sign. If the original signs of the multiplicand and the multiplier were different, the sign of the obtained result must be inverted. The method is intuitive and rather does not require any additional illustration with example. Therefore, we just present its implementation in the assembly code of the 8051 processor.

Implementation in code:

- input: A – multiplicand, B – multiplier,
- output: B – higher byte of result, A – lower byte of result,
- exemplary value: $11111101_{2's}$ * $00000011_{2's}$.

```
 1    ;***************************************************************************
      ;
 2    ;* Multiplication of 2's numbers *
 3    ;* Sign change method *
 4    ;***************************************************************************
      ;
 5    00FD            n EQU 11111101B              ;-3 U2
 6    0003            m EQU 00000011B              ;+3 U2
 7
 8    0000: 74 FD     MOV A,#n                     ;multiplicand
 9    0002: 75 F0 03  MOV B,#m                     ;multiplier
10    0005: 12 00 0A  LCALL _2SMULSIGNCHANGE
11                                                 ;result in {B,A}
12    0008: 80 FE     STOP: SJMP STOP
13    ;--------------------------------------------------------------------------
14    000A:           _2SMULSIGNCHANGE:
15    000A: C0 E0     PUSH ACC
16    000C: 65 F0     XRL A,B
```

```
17   000E: A2 E7        MOV C,ACC.7
18   0010: 92 D5        MOV PSW.5,C              ;1 if signs are not the same
19   0012: D0 E0        POP ACC
20   0014: 30 E7 02     JNB ACC.7,POSITIVE_N
21   0017: F4           CPL A
22   0018: 04           INC A
23   0019:              POSITIVE_N:
24   0019: 30 F7 05     JNB B.7,POSITIVE_M
25   001C: 63 F0 FF     XRL B,#0FFh
26   001F: 05 F0        INC B
27   0021:              POSITIVE_M:
28   0021: A4           MUL AB
29   0022: 30 D5 0A     JNB PSW.5,END
30   0025: F4           CPL A
31   0026: 24 01        ADD A,#1
32   0028: C5 F0        XCH A,B
33   002A: F4           CPL A
34   002B: 34 00        ADDC A,#0
35   002D: C5 F0        XCH A,B
36   002F:              END:
37   002F: 22           RET
38   ;--- end of file ---
```

Multiplication of two 2's numbers – Robertson's method – to discuss the details of this algorithm, we will represent the number B in 2's in a slightly different way. Well, if we notice that (3.1):

$$B_{2's} = -b_{n-1} \cdot 2^{n-1} + b_{n-2} \cdot 2^{n-2} + \dots + b_1 \cdot 2 + b_0 + b_{-1} \cdot 2^{-1} + \dots + b_{-m} \cdot 2^{-m}$$

$$= -b_{n-1} \cdot 2^{n-1} + \sum_{i=-m}^{n-2} b_i \cdot 2^i = -b_{n-1} \cdot 2^{n-1} + \tilde{B}$$

$$(3.1)$$

then we obtain Eq. (3.2):

$$A_{2's} \cdot B_{2's} = A_{2's} \cdot \tilde{B} - b_{n-1} \cdot A_{2's} \cdot 2^{n-1} = \text{pseudoproduct-correction} \qquad (3.2)$$

Here we treat the 2's numbers as BIN numbers, except that the highest bit of multiplier is temporarily removed. The result of the multiplication is a pseudo product needing adjusting by subtracting 'correction' value. The next example illustrates the details of the procedure.

Example 3.24 Multiplication of two 2's numbers with Robertson's method:

a)

$$(+7)_{DEC}$$
$$*\ (+5)_{DEC}$$
$$+35_{DEC}$$

```
            0  0  1  1  1 = A₂'s
       *  0̸  0  1  0  1 = B̃
    0  0  0  0  1  1  1
    0  0  0  0  0  0
 +  0  0  1  1  1  1
    0  1  0  0  0  1  1  pseudoproduct
 -  0  0  0  0  0  0  0  correction
    0  1  0  0  0  1  1  2's
```

b)

$$(+5)_{DEC}$$
$$*\ (-6)_{DEC}$$
$$-30_{DEC}$$

```
            0  1  0  1 = A₂'s
       *  1̸  0  1  0 = B̃
    0  0  0  0  0  0
    0  0  1  0  1
 +  0  0  0  0
    0  0  0  1  0  1  0  pseudoproduct
 -  0  1  0  1  0  0  0  correction
    1  1  0  0  0  1  0  2's
```

c)

$$(-4)_{DEC}$$
$$*\ (+5)_{DEC}$$
$$-20_{DEC}$$

```
            1  1  0  0 = A₂'s
       *  0̸  1  0  1 = B̃
    1  1  1  1  0  0
    0  0  0  0  0
 +  1  1  0  0
    1  1  0  1  1  0  0  pseudoproduct
 -  0  0  0  0  0  0  0  correction
    1  1  0  1  1  0  0  2's
```

d)

$$(-7)_{DEC}$$
$$*\ (-3)_{DEC}$$
$$+21_{DEC}$$

```
            1  0  0  1 = A₂'s
       *  1̸  1  0  1 = B̃
    1  1  1  0  0  1
    0  0  0  0  0
 +  1  0  0  1
    1  0  1  1  1  0  1  pseudoproduct
 -  1  0  0  1  0  0  0  correction
    0  0  1  0  1  0  1  2's
```

The properties of the algorithm give rise to interesting practical indications. Multiplication of two positive numbers by Robertson's algorithm is carried out in the same way as for BIN numbers (see Example 3.24a). In the case of numbers with different signs, it is convenient to take a number with a positive sign as the multiplier and a number with a negative sign as the multiplicand. A disadvantage of the algorithm, which makes its implementation difficult in software, is the necessity to duplicate the sign bit of all the components of the partial sum. Therefore, the MUL instruction of the processor cannot be used, because the processor fills in the missing values with zeros. In the examples given, the duplicated sign bits are underlined to distinguish them. This drawback does not occur in the algorithms of duplicated sign, Booth's algorithm [Booth 1950] or the method proposed by the author of this book [Gryś 2011], called the two-corrections method. They can be implemented in relatively simple embedded systems equipped with 8/16-bit microprocessors, which instruction set architecture do not include a signed multiplication. Representatives of this group of micro-processors include 8051, Freescale 68HC08 and HCS12, Microchip (formerly Atmel) ATtiny, Zilog eZ8core!, STMicroelectronics ST7, Microchip PIC16/18, Texas Instruments MSP430, NEC 78K0S/0R and others. A careful analysis of the machine code generated by compilers of high-level languages, such as C, BASIC, shows that the above methods are commonly used. For example, the Booth method is recommended by Microchip for some of its 8/16-bit microcontrollers. The BASCOM compiler for the 8051 family uses the 'sign conversion method'. The C compiler from Keil uses the 'sign extension method'.

Exercise 3.25: Perform multiplication of 2's numbers according to Robertson's method:

a.

$$110.1_{2's}$$
$$*\ \ 00.11_{2's}$$
$$\overline{????_{2's}}$$

b.

$$1001_{2's}$$
$$*\ \ 1100_{2's}$$
$$\overline{????_{2's}}$$

Multiplication of 2's numbers – sign extension method – double the word length of each number before performing the multiplication. If the numbers are of different lengths, then double the length of the longer number is assumed. The sign bit should be replicated (expanded) to all additional positions. Next, data prepared this way are multiplied like BIN numbers. Only the bits equal to the sum of the number of bits of the original multiplier and the multiplicand are considered. The remaining (higher) bits of the result must be discarded, as the interpretation of all bits can lead to an incorrect result.

Example 3.25 Multiplication of two 2's numbers with Sign extension method:

a)

$$(-4)_{DEC}$$
$$*\ (+5)_{DEC}$$
$$\overline{-20_{DEC}}$$

```
        1 1 1 1 1 1 0 0  (2's)
    *   0 0 0 0 0 1 0 1  (2's)
    -------------------
        1 1 1 1 1 1 1 0 0
      0 0 0 0 0 0 0 0
  +   1 1 1 1 1 1 0 0
    0 0 0 0 0 0 0 0
    -------------------
    1 0 0 1 1 1 0 1 1 0 0  (2's)
```

b)

$$(-3)_{DEC}$$
$$*\ (-2)_{DEC}$$
$$\overline{+6_{DEC}}$$

```
            1 1 1 1 0 1  (2's)
        *   1 1 1 1 1 0  (2's)
        -----------------
            0 0 0 0 0 0
          1 1 1 1 0 1
        1 1 1 1 0 1
      1 1 1 1 0 1
  +   1 1 1 1 0 1
    1 1 1 1 0 1
  -----------------------
  1 1 1 0 1 1 0 0 0 1 1 0  (2's)
```

The correctness of the method can be proved for numbers of any length. For the sake of clarity of the argumentation, we shall limit our consideration to the case of 2-bit numbers without a fractional part. Then $A_{2's} = -a_1 \cdot 2 + a_0$ and $B_{2's} = -b_1 \cdot 2 + b_0$. After the sign bits of both numbers, a_1 and b_1 have been replicated to the left, they are symbolically written in the following way:

$$\underline{a_1}\ \ \underline{a_1}\ \ a_1\ \ a_0$$
$$*\ \ \underline{b_1}\ \ \underline{b_1}\ \ b_1\ \ b_0$$
$$\overline{?\ \ \ \ ?\ \ \ \ ?\ \ \ \ ?}$$

Let's use the direct multiplication method for BIN numbers, where the superscript denotes numbers with an extended sign and '...' denotes terms with weights higher than 2^3, which are rejected according to the assumptions of the method (3.3):

$$A^R_{2's} \cdot B^R_{2's} = (a_1 \cdot 2^3 + a_1 \cdot 2^2 + a_1 \cdot 2 + a_0) \cdot (b_1 \cdot 2^3 + b_1 \cdot 2^2 + b_1 \cdot 2 + b_0)$$
$$= ... -(a_1 \cdot b_0 + a_0 \cdot b_1 + 2 \cdot a_1 \cdot b_1) \cdot 2^3 + (a_1 \cdot b_0 + a_0 \cdot b_1 + a_1 \cdot b_1) \cdot 2^2 \quad (3.3)$$
$$+ (a_1 \cdot b_0 + a_0 \cdot b_1) \cdot 2 + a_0 \cdot b_0$$

Let us introduce an additional term that does not change the value of the expression, i.e., $(a_1 \cdot b_0 + a_0 \cdot b_1) \cdot 2 - (a_1 \cdot b_0 + a_0 \cdot b_1) \cdot 2$ and let us group the terms according to the weights of the power of 2 as in Eq. (3.4):

$$A_{2's}^R \cdot B_{2's}^R = -\cancel{a_1} \cdot \cancel{b_1} \cdot \cancel{2}^4 - (a_1 \cdot b_0 + a_0 \cdot b_1) \cdot 2^3 + (a_1 \cdot b_0 + a_0 \cdot b_1) \cdot 2^2$$
$$+ a_1 \cdot b_1 \cdot 2^2 + 2 \cdot (a_1 \cdot b_0 + a_0 \cdot b_1) \cdot 2 - (a_1 \cdot b_0 + a_0 \cdot b_1) + a_0 \cdot b_0$$
$$= -(a_1 \cdot b_0 + a_0 \cdot b_1) \cdot 2^3 + (a_1 \cdot b_0 + a_0 \cdot b_1) \cdot 2^3 + a_1 \cdot b_1 \cdot 2^2 - (a_1 \cdot b_0 + a_0 \cdot b_1) + a_0 \cdot b_0$$
$$= a_1 \cdot b_1 \cdot 2^2 - (a_1 \cdot b_0 + a_0 \cdot b_1) + a_0 \cdot b_0$$

$$(3.4)$$

The result is identical to that expected for 2-bit numbers (3.5):

$$A_{2's} \cdot B_{2's} = (-a_1 \cdot 2 + a_0) \cdot (-b_1 \cdot 2 + b_0)$$
$$= a_1 \cdot b_1 \cdot 2^2 - (a_1 \cdot b_0 + a_0 \cdot b_1) \cdot 2 + a_0 \cdot b_0$$

$$(3.5)$$

The advantage of the algorithm is that it can use the MUL multiplication instruction of the 8051 processor.

Implementation in code:

- input: A – multiplicand, B – multiplier,
- output: B – higher byte of result, A – lower byte of result,
- exemplary value: 111101002's $*$ 111110002's.

```
1       ;********************************************************************
        ;
2       ;* Multiplication of 2's numbers *
3       ;* Sign extension method *
4       ;
        ;********************************************************************
5       00F4            n EQU 11110100B             ;-12 2's
6       00F8            m EQU 11111000B             ;-8 2's
7
8       0000: 74 F4     MOV A,#n                    ;multiplicand
9       0002: 75 F0 F8  MOV B,#m                    ;multiplier
10      0005: 12 00 0A  LCALL _2SMULSIGNEXT
11                                                  ;result in {B,A}
12      0008: 80 FE     STOP: SJMP STOP
13      ;--------------------------------------------------------------------
14      000A:           _2SMULSIGNEXT:
15      000A: F8        MOV R0,A
16      000B: AA F0     MOV R2,B
17      000D: A2 E7     MOV C,ACC.7
18      000F: 92 D5     MOV PSW.5,C
19      0011: 7D 08     MOV R5,#8
20      0013:           LOOP:
21      0013: A2 D5     MOV C,PSW.5
22      0015: 33        RLC A
23      0016: DD FB     DJNZ R5,LOOP
```

24	0018: F9	MOV R1,A	
25	0019: EA	MOV A,R2	
26	001A: A2 E7	MOV C,ACC.7	
27	001C: 92 D5	MOV PSW.5,C	
28	001E: 7C 08	MOV R4,#8	
29	0020:	LOOP1:	
30	0020: A2 D5	MOV C,PSW.5	
31	0022: 33	RLC A	
32	0023: DC FB	DJNZ R4,LOOP1	
33	0025: FB	MOV R3,A	
34	0026: E8	MOV A,R0	
35	0027: 8A F0	MOV B,R2	
36	0029: A4	MUL AB	
37	002A: AD F0	MOV R5,B	
38	002C: FC	MOV R4,A	;R5*R4=R0*R2
39	002D: E9	MOV A,R1	
40	002E: 8A F0	MOV B,R2	
41	0030: A4	MUL AB	;B*A=R2*R1
42	0031: 2D	ADD A,R5	
43	0032: FD	MOV R5,A	
44	0033: E4	CLR A	
45	0034: 35 F0	ADDC A,B	
46	0036: FE	MOV R6,A	;{R6R5R4} bank 0
47	0037: E8	MOV A,R0	
48	0038: 8B F0	MOV B,R3	
49	003A: D2 D3	SETB RS0	;bank 1
50	003C: A4	MUL AB	
51	003D: AD F0	MOV R5,B	
52	003F: FC	MOV R4,A	;R5*R4=R0*R2
53	0040: C2 D3	CLR RS0	;bank 0
54	0042: E9	MOV A,R1	
55	0043: 8B F0	MOV B,R3	
56	0045: D2 D3	SETB RS0	;bank 1
57	0047: A4	MUL AB	;B*A=R2*R1
58	0048: 2D	ADD A,R5	
59	0049: FD	MOV R5,A	
60	004A: E4	CLR A	
61	004B: 35 F0	ADDC A,B	
62	004D: FE	MOV R6,A	;{R6R5R4} bank 1
63	004E: C2 D3	CLR RS0	
64	0050: AB 0E	MOV R3,6+8H	
65	0052: AA 0D	MOV R2,5+8H	
66	0054: A9 0C	MOV R1,4+8H	
67	0056: E9	MOV A,R1	
68	0057: 2D	ADD A,R5	

69	0058: F5 F0	MOV B,A
70	005A: E5 04	MOV A,4H
71	005C: 22	RET
72	;--- end of file ---	

Note the considerable similarity of the code to the 'Multiplication of BIN numbers 2 bytes × 2 bytes' program described in chapter 3.1.

Please note that the sign extension can also be achieved in a simpler way, as presented below:

```
PUSH ACC    ;push number on stack
RLC A       ;rotate left with C bit; if A<0 then C=1
SUBB A,A    ;A-A-C=FF_HEX for A<0 and 00_HEX for A≥0
POP B       ;pop an original number to B register;
            ;a double-wide number with duplicated sign bit is stored in the pair
            ;of registers {B-higher, A-lower}.
```

This can significantly reduce (almost twice) the length of the program code, as shown below. Implementation in code:

- input number: A – multiplier, B – multiplier,
- output number: B – higher byte of result, A – lower byte of result,
- exemplary value: $1111101_{2's} * 00000011_{2's}$.

1	;***
2	;* Multiplication of 2's numbers *
3	;* Sign extension method – fast *
4	;***
5	00FD	n EQU 11111101B	;−3 2's
6	0003	m EQU 00000011B	;+3 2's
7			
8	0000: 74 FD	MOV A,#n	;multiplicand
9	0002: 75 F0 03	MOV B,#m	;multiplier
10	0005: 12 00 0A	LCALL _2SMULSIGNEXTFAST	
11			;result in {B,A}
12	0008: 80 FE	STOP: SJMP STOP	
13	;--
14	000A:	_2SMULSIGNEXTFAST:	
15	000A: F8	MOV R0,A	;n
16	000B: AA F0	MOV R2,B	;m
17	000D: 33	RLC A	
18	000E: 95 E0	SUBB A,ACC	;ext_n
19	0010: F9	MOV R1,A	
20	0011: E5 F0	MOV A,B	

21	0013: 33	RLC A	
22	0014: 95 E0	SUBB A,ACC	;ext_n
23	0016: FB	MOV R3,A	;ext_m
24	0017: E8	MOV A,R0	
25	0018: A4	MUL AB	
26	0019: FC	MOV R4,A	;LSB(n*m)
27	001A: AD F0	MOV R5,B	;MSB(n*m)
28	001C: E8	MOV A,R0	
29	001D: 8B F0	MOV B,R3	
30	001F: A4	MUL AB	
31	0020: FE	MOV R6,A	;LSB(ext_m*n)
32	0021: EA	MOV A,R2	
33	0022: 89 F0	MOV B,R1	
34	0024: A4	MUL AB	;LSB(m*ext_n)
35	0025: 2E	ADD A,R6	
36	0026: 2D	ADD A,R5	
37	0027: F5 F0	MOV B,A	
38	0029: EC	MOV A,R4	
39	002A: 22	RET	
40	;--- end of file ---		

Exercise 3.26: Perform multiplication of 2's numbers according to 'sign extension method':

a.   b.

$$1101_{2's} \qquad 0.11_{2's}$$
$$* \ \ 0011_{2's} \qquad * \ \ 10.0_{2's}$$
$$\overline{\ ????_{2's}\ } \qquad \overline{\ ???_{2's}\ }$$

Multiplication of 2's numbers – Booth method – the operation of the algorithm can be represented as follows:

1. Clear the higher part of the result and the carry bit.
2. Assign a multiplier to the lower part of the result.
3a. If the previous lowest multiplier bit (shifted to the carry bit) is one, add the multiplier to the higher part of the result.
3b. If the current lowest multiplier bit is one, subtract the multiplier from the higher part of the result.
4. Move the lower part of the result/multiplier to the right, the outgoing bit is hold in carry bit.
5. Shift to the right the higher part of the result with the sign bit unchanged (the highest bit), the outgoing bit is written into the position of the higher bit of the lower part of the result/multiplier.

6. Repeat from step 3 for all bits of the multiplier.

The given rules are used in Example 3.26.

Example 3.26 Multiplication of two 2's numbers with Booth's method:

a.

$$1101_{2's}$$
$$* \quad 0011_{2's}$$
$$\overline{\qquad 0000001\{1\underline{0}\}}$$
$$- \quad 1101$$
$$\overline{0011001\{1\underline{0}\}\rightarrow}$$
$$0001100\{1\underline{1}\}\rightarrow$$

$$+ \quad \begin{cases} 0000110\{0\underline{1}\} \\ 1101 \end{cases}$$

$$\overline{1101110\{0\underline{1}\}\rightarrow}$$
$$1110111\{0\underline{0}\}\rightarrow$$
$$\dotfill$$
$$1111011_{2's}$$

b.

$$101_{2's}$$
$$* \quad 110_{2's}$$
$$\overline{\qquad 00011\{0\underline{0}\}\rightarrow}$$

$$- \quad \begin{cases} 00001\{1\underline{0}\} \\ 101 \end{cases}$$

$$\overline{01101\{1\underline{0}\}\rightarrow}$$
$$00110\{1\underline{1}\}\rightarrow$$
$$\dotfill$$
$$0001101_{2's}$$

In this example shown, the multiplier bits are successively replaced by the bits of the lower part of the result by right shifting the multiplication result. The pair of bits, i.e. the current and the previous multiplier bit (stored in the carry bit), is enclosed in brackets {...}, with the carry bit additionally underlined. You can see that the algorithm works by repeatedly shifting (the number of times from 3 to 6 is equal to the number of multiplier bits) the multiplication result, originally composed of zeros and a multiplier, and adding the multiplier to the older part of the result if we have {01}, or subtracting the multiplier from the older part of the result if we have {10}. A modification of the method, known as 'radix-4 Both', was presented in [McSorley 1961]. There are also variants of it for numbers in the complement code up to 1, or modulo 2n – 1 [Efstathiou 2000]. Other authors propose to take higher values of the system basis p = 4 and p = 8 [Cherkauer 1996] and p = 32 or p = 256 [Seidel 2001]. All these variations are based on bit group analysis just like the original method [Booth 1950], whose derivation we will cite after [Pochopień 2012]. Let us represent one of the numbers in the form (3.6):

$$B_{2's} = -b_{n-1}\cdot2^{n-1} + b_{n-2}\cdot2^{n-2} + \ldots + b_1\cdot2 + b_0 + b_{-1}\cdot2^{-1} + \ldots + b_{-m}\cdot2^{-m}$$
$$(3.6)$$

then (3.7)

$$A_{2's} \cdot B_{2's} = A_{U2} \cdot (-b_{n-1} \cdot 2^{n-1} + b_{n-2} \cdot 2^{n-2} + \ldots + b_1 \cdot 2$$
$$+ b_0 + b_{-1} \cdot 2^{-1} + \ldots + b_{-m} \cdot 2^{-m}) \tag{3.7}$$

From the below observation (3.8):

$$2^1 \cdot 2^i = 2^i + 2^i = 2^{i+1} \tag{3.8}$$

and thus (3.9):

$$2^i = 2^{i+1} - 2^i \tag{3.9}$$

Let's replace 2^i by the term $(2^{i+1} - 2^{-i})$ in the expression for the result of multiplication. We obtain (3.10):

$$A_{2's} \cdot B_{2's} = A_{2's} \cdot \left(\begin{array}{c} - b_{n-1} \cdot (2^n - 2^{n-1}) + b_{n-2} \cdot (2^{n-1} - 2^{n-2}) + b_{n-3} \\ \cdot (2^{n-2} - 2^{n-3}) + \ldots + b_{-m} \cdot (2^{-m+1} - 2^{-m}) \end{array} \right) \tag{3.10}$$

or its equivalent form (3.11):

$$A_{2's} \cdot B_{2's} = A_{2's} \cdot (2^{n-1} \cdot (-b_{n-1} + b_{n-2}) + 2^{n-2} \cdot (-b_{n-2} + b_{n-3})$$
$$+ \ldots + 2^{-m} \cdot (-b_m + b_{-m-1})) \tag{3.11}$$

Now we have (3.12):

$$A_{2's} \cdot B_{2's} = (2^{n-1} \cdot (-b_{n-1} + b_{n-2}) \cdot A_{U2} + 2^{n-2} \cdot (-b_{n-2} + b_{n-3}) \cdot$$
$$A_{U2} + \ldots + 2^{-m} \cdot (-b_m + b_{-m-1}) \cdot A_{2's}) \tag{3.12}$$

and finally (3.13):

$$A_{2's} \cdot B_{2's} = \sum_{i=n-1}^{-m} 2^i \cdot (-b_i + b_{i-1}) \cdot A_{2's} \tag{3.13}$$

The multiplication is done by repeated summation of partial products, i.e. the first number A by the weighted difference of two adjacent bits of the multiplier. Depending on their combination, the term $(b_i - b_{i-1})$ vanishes for the same values of bits {00} or {11} is added to the current value of result for {01} or subtracted for the combination {10}. It is worth remembering that the algorithm has one limitation, namely, it returns an incorrect multiplication result for multiplicand of value $10\ldots0_{2's}$!

Implementation in code:

- input: A – multiplicand, B – multiplier,
- output: B – higher byte of result, A – lower byte of result, OV – out of the range caused by using prohibited multiplicand value 10 ... 0B,
- exemplary value: $11111101_{2's} * 00000011_{2's}$.

```
 I     ;***************************************************************
 2     ;* Multiplication of 2's numbers *
 3     ;* Booth method *
 4     ;***************************************************************
 5     00FD            n EQU 11111101B              ;−3 2's
 6     0003            m EQU 00000011B              ;+3 2's
 7
 8     0000: 74 FD     MOV A,#n                     ;multiplicand
 9     0002: 75 F0 03  MOV B,#m                     ;multiplier
10     0005: 12 00 0A  LCALL _2SMULBOOTH
11                                                  ;result in {B,A}
12     0008: 80 FE     STOP: SJMP STOP
13     ;-------------------------------------------------------------------
14     000A:           _2SMULBOOTH:
15     000A: B4 80 03  CJNE A,#80h,MULTIPLY
16     000D: D2 D2     SETB OV
17     000F: 22        RET
18     0010:           MULTIPLY:
19     0010: 7A 08     MOV R2,#8
20     0012: C3        CLR C
21     0013: F8        MOV R0,A
22     0014: E4        CLR A
23     0015:           LOOP:
24     0015: 50 01     JNC SKIP
25     0017: 28        ADD A,R0
26     0018:           SKIP:
27     0018: 30 F0 02  JNB B.0,SKIP1
28     001B: C3        CLR C
29     001C: 98        SUBB A,R0
30     001D:           SKIP1:
31     001D: A2 E7     MOV C,ACC.7
32     001F: 13        RRC A
33     0020: C5 F0     XCH A,B
34     0022: 13        RRC A
35     0023: C5 F0     XCH A,B
36     0025: DA EE     DJNZ R2,LOOP
37
```

38	0027: C5 F0	XCH A,B
39	0029: 22	RET
40	;--- end of file ---	

Exercise 3.27: Perform multiplication of 2's numbers according to Booth method:

a.

$$1101_{2's}$$
$$* \ 0011_{2's}$$
$$\overline{????_{2's}}$$

b.

$$011_{2's}$$
$$* \ 100_{2's}$$
$$\overline{???_{2's}}$$

Multiplication of 2's numbers – two corrections method (proposed by the author of this book) – let us start by deriving the theoretical basis of how the method works. Let A and B denote sign numbers in the two's complement code, consisting of n bits in the integer part and m in the fractional part. The number of bits of A and B in the general case may differ, hence $A(n1, m1)$ and $B(n2, m2)$. The values of the numbers are defined as follows (3.14):

$$A_{2's} = -a_{n1-1} \cdot 2^{n1-1} + \sum_{i=-m1}^{n1-2} a_i \cdot 2^i = -a_{n1-1} \cdot 2^{n1-1} + \tilde{A}$$

$$B_{2's} = -b_{n2-1} \cdot 2^{n2-1} + \sum_{i=-m2}^{n2-2} b_i \cdot 2^i = -b_{n2-1} \cdot 2^{n2-1} + \tilde{B} \qquad (3.14)$$

The symbols $\tilde{A}$ and $\tilde{B}$ represent the positive component of the numbers A and B. Using the above symbols, the product of these numbers can be written (3.15):

$$
\begin{aligned}
A_{2's} \cdot B_{2's} &= (-a_{n1-1} \cdot 2^{n1-1} + \tilde{A}) \cdot (-b_{n2-1} \cdot 2^{n2-1} + \tilde{B}) \\
&= a_{n1-1} \cdot 2^{n1-1} \cdot b_{n2-1} \cdot 2^{n2-1} + a_{n1-1} \cdot 2^{n1-1} \cdot \tilde{B} + b_{n2-1} \cdot 2^{n2-1} \cdot \tilde{A} \\
&\quad + \tilde{A} \cdot \tilde{B} - 2 \cdot b_{n2-1} \cdot 2^{n2-1} \cdot \tilde{A} - 2 \cdot a_{n1-1} \cdot 2^{n1-1} \cdot \tilde{B} \qquad (3.15) \\
&= A_{BIN} \cdot B_{BIN} - b_{n2-1} \cdot 2^{n2} \cdot \tilde{A} - a_{n1-1} \cdot 2^{n1} \cdot \tilde{B} \\
&= \text{pseudoproduct} - (\text{correction\_A} + \text{correction\_B})
\end{aligned}
$$

The multiplication result is calculated in two steps. In the first step, the multiplied numbers are treated as binary unsigned numbers. Such a preliminary result of multiplication is called, like in Robertson's method, a 'pseudo product'. If the multiplied numbers are positive signs, it becomes the multiplication result and operation is completed. In other cases, one or two corrections called 'correction_A' and/or 'correction_B' respectively are necessary. These are calculated as the product of the three components (3.16):

$$correction\_A = b_{n2-1}\cdot 2^{n2}\cdot\tilde{A}$$
$$correction\_B = a_{n1-1}\cdot 2^{n1}\cdot\tilde{B} \tag{3.16}$$

The operation of the method is illustrated by a numerical example.

Example 3.27: Determine the product of the numbers A and B expressed in terms of the least number of bits.

a) A=-8$_{DEC}$ B=-3$_{DEC}$

Solution

A =1000$_{2's}$ B=101$_{2's}$

n$_1$=4 m$_1$=0   n$_2$=3 m$_2$=0

```
          1 0 0 0 = A 2's
     *      1 0 1 = B 2's
          1 0 0 0
        0 0 0 0
   + 1 0 0 0
   ----------------
     1 0 1 0 0 0 = pseudoproduct
   - 0 0 0 0 0 0 = correction_A
   - 0 1 0 0 0 0 = correction_B
     0 1 1 0 0 0 2's
```

b) A =-3.5$_{DEC}$ B=+1.5$_{DEC}$

Solution

A =100.1$_{2's}$ B=01.10$_{2's}$

n$_1$=3 m$_1$=1   n$_2$=2 m$_2$=2

```
           1 0 0. 1 = A 2's
     *       0 1. 1 0 = B 2's
           0 0 0 0
         1 0 0 1
   +   1 0 0 1
   -----------------
     0 1 1 0. 1 1 0 = pseudoproduct
   - 0 0 0 0. 0 0 0 = correction_A
   - 1 1 0 0. 0 0 0 = correction_B
     1 0 1 0. 1 1 0 2's
```

In practical applications the multiplied numbers are usually stored in the processor memory as numbers of the same format, so $n_1 = n_2 = n$ and $m_1 = m_2 = m$. The equation for their multiplication then takes a simplified form (3.17):

$$A_{2's}\cdot B_{2's} = (-a_{n-1}\cdot 2^{n-1} + \tilde{A})\cdot(-b_{n-1}\cdot 2^{n-1} + \tilde{B}) = A_{BIN}\cdot B_{BIN} - 2^n$$
$$\cdot(b_{n-1}\cdot\tilde{A} + a_{n-1}\cdot\tilde{B}) \tag{3.17}$$

The required number of bits of the multiplication result is equal to $2*(n + m)$ if each of the multiplied numbers is of size $n + m$. The correction of pseudo product (conditional subtraction dependent on the signs of A and B) are performed only on the higher part of the 'pseudo product', since the last term $b_{n-1}\cdot\tilde{A} + a_{n-1}\cdot\tilde{B}$ is scaled by a factor of 2^n. Another possible reduction in the complexity of the algorithm requires consideration of combinations of number signs. As noted earlier, if A and B are positive numbers the pseudo random corrections are not needed. On the other hand, if the numbers are of different signs, then one of the two corrections must be performed. The following features of a real processor: fixed length of registers word, automatic filling with zero of the leading bits of registers (conventionally from the left), allow to further reduction of the complexity of the method, in the context of its efficient implementation. Note additionally that the highest bit of a positive number in the code of additions to 2 is zero, then $\tilde{A} = A$ and $\tilde{B} = B$. As a consequence of this, they can be replaced by A and B in the expressions for 'correction_A' and 'correction_B' as (3.18):

$$A_{2's} \cdot B_{2's} = A_{BIN} \cdot B_{BIN} - 2^n \cdot (b_{n-1} \cdot A + a_{n-1} \cdot B)_{\text{with some assumptions}} \qquad (3.18)$$

This situation is shown in Example 3.28, which highlights the bits automatically filled with zeros by the processor.

Example 3.28: Multiply the numbers A and B assuming $n = 3$ and $m = 1$. Let $A = -3.5_{DEC}$ and $B = 3.0_{DEC}$.

Solution

$A = 100.1_{2's}$ $B = 011.0_{2's}$

a) according to Eq. (3.17)

```
          1  0  0.  1 = A₂'ₛ
       *  0  1  1.  0 = B₂'ₛ
       ─────────────────
          0  0  0  0
       1  0  0  1
    1  0  0  1
 +  0  0  0  0
 ──────────────────────
 0  0  1  1  0  1.  1  0 = pseudoproduct
-0  0  0  0  0  0.  0  0 = correction_A
-0  1  1  0  0  0.  0  0 = correction_B
 ──────────────────────
 1  1  0  1  0  1.  1  0 ₂'ₛ
```

b) according to Eq. (3.18)

```
          1  0  0.  1 = A₂'ₛ
       *  0  1  1.  0 = B₂'ₛ
       ─────────────────
          0  0  0  0
       1  0  0  1
    1  0  0  1
 +  0  0  0  0
 ──────────────────────
 0  0  1  1  0  1.  1  0 = pseudoproduct
-0  0  0  0  0  0.  0  0 = correction _A
-0  1  1  0  0  0.  0  0 = correction _B
 ──────────────────────
 1  1  0  1  0  1.  1  0 ₂'ₛ
```

Performing calculations according to both relations (3.17) and (3.18), we obtain an identical result, as expected. The above modification is valid also in the case when both numbers are negative. This can be proved by the following observation. Namely, if the borrowing bit from the highest position of the result is discarded then it is true that (3.19)

$$\text{bit} - 0 - 0 = \text{bit} - 1 - 1 = \text{bit} \qquad (3.19)$$

Example 3.29 shows the principle of the proposed method precisely in the case of both negative numbers. As before, the correct result is obtained both by calculating the product according to formula (3.17) and (3.18).

Example 3.29: Multiply the numbers A and B assuming $n = 4$ and $m = 0$. Let $A = -6_{DEC}$ and $B = -3_{DEC}$.

Solution

$A = 1010_{2's}$ $B = 1101_{2's}$

a) according to Eq. (3.17)

$$
\begin{array}{r}
1\ 0\ 1\ 0 = A_{2's} \\
*\ \ 1\ 1\ 0\ 1 = B_{2's} \\
\hline
1\ 0\ 1\ 0 \\
0\ 0\ 0\ 0 \\
1\ 0\ 1\ 0 \\
+\ 1\ 0\ 1\ 0 \\
\hline
1\ 0\ 0\ 0\ 0\ 0\ 1\ 0 = \text{pseudoproduct} \\
-\ \underline{0}\ 0\ 1\ 0\ 0\ 0\ 0\ 0 = \text{correction\_A} \\
-\ \underline{0}\ 1\ 0\ 1\ 0\ 0\ 0\ 0 = \text{correction\_B} \\
\hline
0\ 0\ 0\ 1\ 0\ 0\ 1\ 0_{2's}
\end{array}
$$

b) according to Eq. (3.18)

$$
\begin{array}{r}
1\ 0\ 1\ 0 = A_{2's} \\
*\ \ 1\ 1\ 0\ 1 = B_{2's} \\
\hline
1\ 0\ 1\ 0 \\
0\ 0\ 0\ 0 \\
1\ 0\ 1\ 0 \\
+\ 1\ 0\ 1\ 0 \\
\hline
1\ 0\ 0\ 0\ 0\ 0\ 1\ 0 = \text{pseudoproduct} \\
-\ 1\ 0\ 1\ 0\ 0\ 0\ 0\ 0 = \text{correction\_A} \\
-\ 1\ 1\ 0\ 1\ 0\ 0\ 0\ 0 = \text{correction\_B} \\
\hline
0\ 0\ 0\ 1\ 0\ 0\ 1\ 0_{2's}
\end{array}
$$

Solution

Implementation in code:

- input: A – multiplicand, B – multiplier,
- output: B – higher byte of result, A – lower byte of result,
- exemplary value: $11111101_{2's} * 00000011_{2's}$.

```
1       ;***************************************************************
2       ;* Multiplication of 2's numbers *
3       ;* Two corrections method (proposed by Grys) *
4       ;***************************************************************
5       00FD                    n EQU 11111101B              ;-3 2's
6       0003                    m EQU 00000011B              ;+3 2's
7
8       0000: 74 FD             MOV A,#n                     ;multiplicand
9       0002: 75 F0 03          MOV B,#m                     ;multiplier
```

```
10     0005: 12 00 0A              LCALL _2SMULGRYS
11                                                          ;result in {B,A}
12     0008: 80 FE                 STOP: SJMP STOP
13     ;----------------------------------------------------------------------------------------
14     000A:                       _2SMULGRYS:
15     000A: F5 20                 MOV 20h,A
16     000C: 85 F0 21              MOV 21h,B
17     000F: A4                    MUL AB                   ;C=0
18     0010: C5 F0                 XCH A,B
19     0012: 30 07 02              JNB 20h.7,POSITIVE_N
20     0015: 95 21                 SUBB A,21h
21     0017:                       POSITIVE_N:
22     0017: 30 0F 03              JNB 21h.7,POSITIVE_M
23     001A: C3                    CLR C
24     001B: 95 20                 SUBB A,20h
25     001D:                       POSITIVE_M:
26     001D: C5 F0                 XCH A,B
27     001F: 22                    RET
28     ;--- end of file ---
```

By comparing the proposed method with other multiplication methods, based on their properties and code analysis, its advantages and disadvantages can be identified. The detailed discussion on comparison was presented in [Gryś 2011]. Here, we recall the general features. Advantages of the 'two corrections' method and its implementation in assembly code as presented above are as follows:

- It works correctly for A = 10...0 unlike Booth's method.
- Does not work in a loop, hence execution time does not depend on word length.
- The smallest code size (the smallest occupation of program memory).
- The shortest execution time (the fastest method).
- Smaller register occupancy compared to the extended sign method.

Disadvantages of the proposed 'two correction' method and its implementation:

- for negative numbers one or two corrections of the result are required;
- the processor instruction list must include the unsigned multiplication operation (otherwise it must be emulated by software);
- execution time (measured in processor cycles) is variable and depends on the sign combinations of the numbers, contrary to the extended sign method.

Division of two 2's numbers is complicated. An example of one of possible algorithm is discussed in [Pochopień 2012]. In practice, it is more convenient to convert negative 2's numbers into their positive counterparts, perform the division as BIN numbers, and in the case of different original signs of the numbers finally convert the result back into a negative 2's number. The relevant algorithms were presented in the previous chapters 2.4 and 3.1.1.

Exercise 3.28*: Write a subroutine for division of 2's numbers by changing the signs.

3.3 NONLINEAR FUNCTIONS

In many applications not only related to the academic research or engineering, we need using various functions, e.g. trigonometric and hyperbolic functions, exponent, logarithms, power, square and less often n-th order root, etc. Let's consider some examples of a nonlinear function lending itself to a real-world scenario. In what situations we need, e.g., trigonometric function as well? Please think about drawing a circle or rotating geometrical figures with assumed angle. It is presented in computer graphics and animation. The hyperbolic cosine is a function used to describe mathematically the shape of a dangling electric power line or rope suspended at the ends. The hyperbolic functions are used to describe the so called 'hyperbolic motion' in relativistic physics. Logarithm and exponent are useful in electrical engineering during analysis of the DC circuits in transient state or AC circuits in electrical power production and distribution system. An example in biology is exponential growth of a bacterial colony. The logarithmic scale is very common in techniques like: acoustics – sound level in dB, pH for acidity. Richter magnitude scale and moment magnitude scale for strength of earthquakes and movement in the Earth are based on logarithm also. The exponentiation, particularly raising to power is very common. Good examples are kinetic energy, moment of inertia formulas taught even at early education level. What about the square root? It is used in electrical engineering again to express the ration between amplitudes of voltages in one- and three-phase electrical networks. We can see it working with geometric and harmonic mean, staying familiar with basic physic at school, e.g. pendulum – dependence of the oscillation period on the amplitude, also it is used in the logistics management and many, many others. There are no doubts that after a moment's of thought probably anybody can propose different examples of functions applied in many disciplines of science, engineering, education and normal life. These above mentioned functions are usually strongly nonlinear and cannot be calculated directly by simple processor with limited arithmetic capabilities even if we assume that we are able to work with fractions as shown in previous material discussed in this chapter. Nonlinear functions can be computed by its approximation engaged four basic operations only, lookup table, iterative operations in the loop or conditional pieces of code related to the value of input argument – piecewise approximation.

For integer input numbers, required function can be approximated by pair of points {input number, function value} and tabled. As an example of this fundamental method, the square root function was selected in below listing. The expected values are located in table section (after label named TABLE:) after simple code. For simplicity only initial part of table was prepared for numbers from subrange <0,5>. There is no problem extending table for full scale <0,255>.

Implementation in code:

- input: A – number (must be integer),
- output: R0 – integer part of result, A – fractional part of result,
- exemplary value: $sqrt(00000010_{BIN}) = 00000001_{P\text{-}BCD}$, $01000001_{P\text{-}BCD}$.

```
 1      ;************************************************************
 2      ;* Square root by LUT *
 3      ;************************************************************
 4      0003            n EQU 00000011B              ;+3 BIN
 5
 6      0000: 74 03     MOV A,#n
 7      0002: 12 00 07  LCALL SQRT_LUT
 8                                                   ;result in R0-integer
 9                                                   ;R1-fraction
10
11      0005: 80 FE     STOP: SJMP STOP
12      ;-----------------------------------------------------------------
13      0007:           SQRT_LUT:
14      0007: 90 00 15  MOV DPTR,#TABLE
15      000A: 23        RL A                         ;multiply by 2 (address
                                                      adjusted)
16      000B: F5 F0     MOV B,A
17      000D: 93        MOVC A,@A+DPTR
18      000E: F8        MOV R0,A
19      000F: E5 F0     MOV A,B
20      0011: 04        INC A
21      0012: 93        MOVC A,@A+DPTR
22      0013: F9        MOV R1,A
23      0014: 22        RET
24      0015:           TABLE:;two bytes in P-BCD
                          format, e.g. sqrt(2)=1.41
25      0015: 00 00     DB 00000000B,00000000B       ;for 0
```

26	0017: 01 00	DB 00000001B,00000000B	;for 1
27	0019: 01 41	DB 00000001B,01000001B	;for 2
28	001B: 01 73	DB 00000001B,01110011B	;for 3
29	001D: 02 00	DB 00000010B,00000000B	;for 4
30	001F: 02 24	DB 00000010B,00100100B	;for 5
31	; etc.		
32	; DB...		
33	;--- end of file ---		

Exercise 3.29: What should be modified in above listing if we need result as BIN numbers or greater fraction precision?

Different approach for square root function was applied below using known observation. If the result can be limited to integer part only, e.g. n = 11 and sqrt(n) = 3 or n = 26 and then sqrt(n) = 5, we can construct rolling algorithm. We are summing up only odd numbers starting from 1 like this i = 1, 3, 5, 7, ... and sum = 1 + 3 + 5 + 7 + ... until we get first time the condition sum > n. Hence, integer approximation of sqrt(number) = i − 1/2. For easier understanding this rule and implementation in assembly code the some simple calculations are provided in Table 3.3.

Table 3.3 Explanation How to Estimate Square Root of n

n	sqrt(n)	Sum	last i	Approx. of sqrt(n) = (i − 1)/2
0	0.00	(1 = 1) > 0	1	0
1	1.00	(1 + 3 = 4) > 1	3	1
2	1.41	(1 + 3 = 4) > 2	3	1
3	1.73	(1 + 3 = 4) > 3	3	1
4	2.00	(1 + 3 + 5 = 9) > 4	5	2
...				
11	3.32	(1 + 3 + 5 + 7 = 16) > 11	7	3
...				
16	4.00	(1 + 3 + 5 + 7 + 9 = 23) > 16	9	4

Implementation in code:

- input: A − number (must be integer),
- output: A − integer part of result,
- exemplary value: sqrt(00001011_{BIN}) = 00000011_{BIN}.

| 1 | ;*** |
| 2 | ;* Square root by addition * |

```
3       ;***************************************************************
4       000B                    n EQU 00001011B          ;+11 BIN
5
6       0000: 74 0B             MOV A,#n
7       0002: 12 00 07          LCALL SQRT_ITER
8                                                        ;result in A-integer part only
9                                                        ;truncation
10      0005: 80 FE             STOP: SJMP STOP
11      ;-------------------------------------------------------------------
12      0007:                   SQRT_ITER:
13      0007: F8                MOV R0,A
14      0008: 74 FF             MOV A,#0FFH              ;i=-1
15      000A: 75 F0 00          MOV B,#0                 ;sum
16      000D: C0 E0             PUSH ACC
17      000F:                   LOOP:
18      000F: D0 E0             POP ACC
19      0011: 04                INC A
20      0012: 04                INC A                    ;1..3..5..etc
21      0013: C0 E0             PUSH ACC
22      0015: 25 F0             ADD A,B                  ;sum=sum+i
23      0017: 60 09             JZ SKIP                  ;for n>225
24      0019: F5 F0             MOV B,A
25      001B: B5 00 02          CJNE A,0,NOT_EQUAL       ;R0 has address 0
26      001E: 80 EF             SJMP LOOP
27      0020:                   NOT_EQUAL:
28      0020: 40 ED             JC LOOP
29      0022: D0 E0 SKIP:       POP ACC
30      0024: 14                DEC A
31      0025: 03                RR A                     ;sqrt(n)=(i-1)/2
32      0026: 22                RET
33      ;--- end of file ---
```

There exist other methods for square approximation, e.g. based on initial estimate, Heron's, Bakhashali, exponential, digit-by-digit method or Taylor series. So far we have shortly discussed simple methods of evaluation of square root as an example of nonlinear functions. These methods were adapted for fixed-point format and limited to integer argument only. Some further questions may arise here even if we continue considerations for fixed-point format. How to deal with input argument being a real number? What about the other functions except square root, e.g. trigonometric or logarithms? Do we really have to look for individual methods of approximation or is there any universal technique for precise and quick function evaluation? Luckily there is a way to do that. It is named CORDIC proposed many years ago by Jack Volder [Volder 1959] and commonly applied nowadays. The CORDIC abbreviation is from 'coordinate rotation digital computer'.

The sine and cosine of an angle θ are determined by rotating the unit vector [1, 0] through decreasing angles until the cumulative sum of the rotation angles equals the input angle. The x and y Cartesian components of the rotated vector then correspond, respectively, to the cosine and sine of θ. Inversely, the angle of a vector [x, y], corresponding to arctangent (y/x), is determined by rotating [x, y] through successively decreasing angles to obtain the unit vector [1, 0]. The cumulative sum of the rotation angles gives the angle of the original vector. The CORDIC algorithm can also be used for calculating hyperbolic functions by replacing the successive circular rotations by steps along a hyperbola. Thanks to this idea computers can calculate the following functions: cosine (cos(x)), sine (sin(x)), atan2(y,x), modulus i.e. sqrt(x^2+y^2), arctangent (tan^{-1}(x)), hyperbolic sin (sinh(x)), hyperbolic cosine (cosh(x)) and hyperbolic arctangent (atanh(x)). If needed, the other functions can be evaluated from known identities like below, e.g.:

$$
\begin{aligned}
\tanh(x) &= \sinh(x)/\cosh(x) \\
\coth(x) &= 1/\tanh(x) \\
\text{arccoth}(z) &= \tfrac{1}{2} * \ln((z + 1)/(z - 1)) \\
\ln(x) &= 2 * \text{atanh}((x + 1)/(x - 1)) \\
\log10(x) &= \log 10(e) * \ln(x) = 0.434294482 * \ln(x) \\
\exp(a) &= \sinh(a) + \cosh(a)
\end{aligned}
$$

From the algorithmic point of view, the CORDIC can be seen as a sequence of micro rotations, where the vector XY is rotated by an angle θ expressed in radians. The algorithm foundations will be cited after [Vitali 2017]. Remembering that tan(θ) = sin(θ)/cos(θ) and applying the known in computer graphics the affine transformation for rotation we obtain as following (3.20):

$$
\begin{aligned}
X_{n+1} &= \cos(\theta) * X_n - \sin(\theta) * Y_n = \cos(\theta) * [X_n - \tan(\theta) * Y_n] \\
Y_{n+1} &= \sin(\theta) * X_n + \cos(\theta) * X_n = \cos(\theta) * [\tan(\theta) * X_n + Y_n]
\end{aligned}
$$

$$(3.20)$$

For simplification needed calculations, the rotation angle is chosen so that the tan(θ) coefficient is a power of 2. Therefore the multiplication is replaced with bit shift to the right realized easy by microprocessor instruction. If the components are scaled by F = 1/cos(θ), which is the CORDIC gain, the formula for the rotation is reduced indeed to only bit shifts and additions (3.21):

$$
\begin{aligned}
X_{n+1} * F_n &= [X_n - Y_n/2^n] \\
Y_{n+1} * F_n &= [X_n/2^n + Y_n]
\end{aligned}
$$

$$(3.21)$$

The CORDIC algorithm can work in circular of hyperbolic modes. To keep the text readable, we will show in more details the circular mode only. The elementary rotation angle is then $\theta_n = \text{atan}(2^{-n})$. The corresponding scaling factor for n-th step is $F_n = 1/\cos(\theta_n) = \text{sqrt}(1 + 2^{-2n})$. In the first iteration n = 0, vector is rotated 45° counterclockwise to get the updated vector. Successive iterations rotate the vector in one or the other direction by size-decreasing steps, until the desired angle has been achieved. As only all shifts and adds are done for all assumed iteration, the final output vector components X and Y are scaled by the factor F. It is a multiplication of all $1/F_n$ and its value depends on number of iterations. For most ordinary purposes, 40 iterations (n = 40) is sufficient to obtain the correct result of calculated function to the 10th decimal place and F = 0.607252935008881. The CORDIC algorithm can also be used for calculating hyperbolic functions (sinh, cosh, atanh) by replacing the circular rotations by hyperbolic angles $\text{atanh}(2^{-j})$, where j = 1, 2, 3, ..., n then F = 1.20513635844646.

Because the CORDIC implementation done in assembly language for 8051 CPU would be extensive and probably not readable, only one in this place of this book, we decided to present it in a high-level language, i.e. Matlab/GNU Octave as below. It is a simplified version of a code available in Wikipedia article [Wiki 2022] adapted to above theoretical considerations and symbols.

```
% We compute v = [cos(beta), sin(beta)] (beta in radians)using n iterations.
beta=beta*10000;
% Initialization of tables of constants used by CORDIC need a table
% of arctangents of negative powers of two, in radians:
% angles = atan(2.^-(0:23)); % but we will use approximated values
instead
angles = [7854 4636 2450 1244 624 312 156 78 39 20 10 5 2 1 1];
% and a table of products of scaling factors Fn:
% Fn = cumprod(sqrt(1 + 2.^(-2*(0:23)))) % because 45/2 = 23

Fn = [ 1414 1581 1629 1642 1646 1646 1647 1647];
Fn(9:23)= 1648;

F = 1/Fn(n)*1000;

% Initialize loop variables:
v = [1;0]; % start with two-vector cosine and sine of zero
poweroftwo = 1; % because 2^0=1
angle = angles(1);
% Iterations
for j = 0:n-1;
    if beta < 0
        dir = -1;
```

```
else
   dir = 1;
end
factor = dir * poweroftwo;
R = [1, -factor; factor, 1];
v = R * v; % 2-by-2 matrix multiply
beta = beta - dir * angle; % update the remaining angle
poweroftwo = poweroftwo / 2;
angle = angles(j+2);
end
% Adjust length of output vector to be [cos(beta), sin(beta)]:
v = v * F;
% ------- end of code ----------
```

We performed some test, and in Figure 3.1 we compare values of sin() and cos() functions evaluated using above cod with ideal shapes for n = 8. Some imperfections are visible. Increasing n parameter will increase the precision.

For n > 10 would be hard to see differences because the maximal error related to maximal amplitude value os sin/cos decreases rapidly, as shown in Figure 3.2. It refers to cosine function too.

For further reading on fixed-point arithmetic, we can recommend selected papers and books [Baer 2010, Flores 1962, Hwang 1979, Koren 1993, Kulisch 2012, Mano 1993, Mano 2008, Omondi 1994, Parhami 2010,

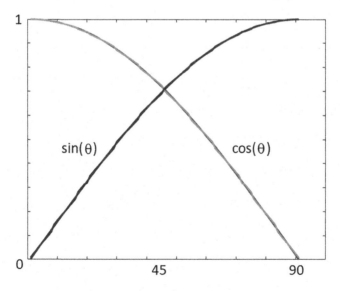

Figure 3.1 Functions avaluated by CORDIC algorithm vs. perfect sin/cos function shapes.

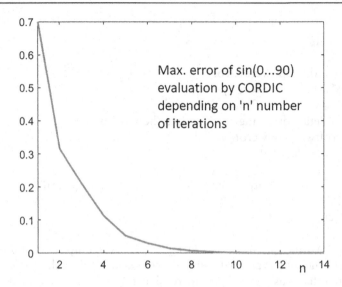

Figure 3.2 Maximal error of sinus evaluation by CORDIC algorithm.

Pochopień 2012, Richards 1955, Schmid 1979 and Swartzlander 2015]. In Chapters 4 and 5, we talk about floating-point format being a very effective way to express real numbers in the 'world of computers'. As we will see the user is practically not limited in number range and precision that sound good and would satisfy even very demanding computer users as microphysics scientists, astronomers and engineers.

Chapter 4

Numbers in Floating-point Format

The number A is written in (n + m) digits with base p, where n is the number of digits of the mantissa M, m is the number of digits of the characteristic (exponent) E (Figure 4.1).

Please note the different format of the above bit field compared to the fixed-point format. The term 'floating-point' emphasizes the possibility of expressing the same value of a number using different combinations of mantissa and exponent, as shown in Example 4.1. In the following section, we will assume p = 2, considering only binary systems.

Example 4.1: Decimal number 2.5_{DEC} expressed as BIN using 8 bits:

- in fixed-point format 0010.1000
- in floating-point format 0.10100E10 = $0.10100*p^{10}$, or 0.0101E011 = $0.0101*p^{011}$, where p = 2 − radix, etc.

4.1 NON-NORMALIZED NUMBERS

From a practical point of view, it is important to adopt a convention for interpreting bit-field values such that signed numbers can be stored. In principle, there is no barrier obstacle to, for example, expressing the mantissa in the 2's complement code and the exponent in the sign-magnitude. If one restricts consideration solely to these two most commonly used forms of writing numbers with sign, four combinations are obtained (4.1):

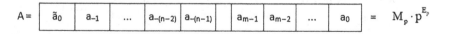

$$A = \boxed{\begin{array}{|c|c|c|c|c|c|c|c|}\hline \tilde{a}_0 & a_{-1} & \cdots & a_{-(n-2)} & a_{-(n-1)} & a_{m-1} & a_{m-2} & \cdots & a_0 \\ \hline\end{array}} = M_p \cdot p^{E_p}$$

Figure 4.1 A floating-point format of number.

DOI: 10.1201/9781003363286-4

$$A = M_{2's} \cdot 2^{E_{SM}} \text{ or } A = M_{2's} \cdot 2^{E_{2's}} \text{ or } M_{SM} \cdot 2^{E_{2's}} \text{ or } M_{SM} \cdot 2^{E_{SM}} \qquad (4.1)$$

Assuming that the mantissa is a fraction and the exponent an integer, the following rules for determining their values can be adopted (4.2):

$$M_{SM} = (-1)^{\tilde{a}_0} \cdot (a_{-1} \cdot 2^{-1} + \ldots + a_{-(n-1)} \cdot 2^{-(n-1)}) = (-1)^{\tilde{a}_0} \cdot \sum_{i=-1}^{-(n-1)} a_i \cdot 2^i \quad (4.2a)$$

$$E_{SM} = (-1)^{a_{m-1}} \cdot (a_{m-2} \cdot 2^{m-2} + \ldots + a_1 \cdot 2 + a_0) = (-1)^{a_{m-1}} \cdot \sum_{i=0}^{m-2} a_i \cdot 2^i \quad (4.2b)$$

$$M_{2's} = -\tilde{a}_0 + a_{-1} \cdot 2^{-1} + \ldots + a_{-(n-1)} \cdot 2^{-(n-1)} = -\tilde{a}_0 + \sum_{i=-1}^{-(n-1)} a_i \cdot 2^{-i} \quad (4.2c)$$

$$E_{2's} = -a_{m-1} \cdot 2^{m-1} + a_{m-2} \cdot 2^{m-2} + \ldots + a_1 \cdot 2 + a_0$$
$$= -a_{m-1} \cdot 2^{m-1} + \sum_{i=0}^{m-2} a_i \cdot 2^i \qquad (4.2d)$$

The mantissa lies in the range $-(1 - 2^{-(n-1)}) \le M_{SM} \le 1 - 2^{-(n-1)}$ for the SM sign-magnitude format and, respectively, $-1 \le M_{2's} \le 1 - 2^{-(n-1)}$ for the 2's notation. Zero has two representations –0 and +0. Let's list the smallest and largest mantissa values:

- the smallest negative: $1.11...1_{SM} = -(1 - 2^{-(n-1)})$ $1.00...0_{2's} = -1$,
- the largest negative: $1.00...0_{SM} = -0$ $1.11...1_{2's} = -2^{-(n-1)}$,
- the smallest positive: $0.00...0_{SM} = +0$ $0.00...02s = +0$,
- the largest positive: $0.11...1_{SM} = 1 - 2^{-(n-1)}$ $0.11...1_{2's} = 1 - 2^{-(n-1)}$.

The exponent lies in the range $-(2^{m-1} - 1) \le E_{SM} \le 2^{m-1} - 1$ for the SM sign-magnitude format and, respectively, $-2^{m-1} \le E_{2's} \le 2^{m-1} - 1$ for 2's notation. Zero has two representations –0 and +0 as previously. Let's list the smallest and largest exponent values:

- the smallest negative: $1\ 11...1_{SM} = -(2^{m-1} - 1)$ $100...0_{2's} = -2^{m-1}$,
- the largest negative: $1\ 00...0_{SM} = -0$ $111...1_{2's} = -1$,
- the smallest positive: $0\ 00...0_{SM} = +0$ $000...0_{2's} = +0$,
- the largest positive: $0\ 11...1_{SM} = 2^{m-1} - 1$ $011...1_{2's} = 2^{m-1} - 1$.

Which form is more beneficial, if we take the proximity of implementation as a criterion for choice? To answer this question, it is important to note that the operations of addition and subtraction of numbers in fixed-point format are more easily performed for 2's format, while multiplication and division are performed for SM format. It is also important to note that when we want to apply each of the four basic arithmetic operations on two floating-point numbers, there will be a need to add, subtract, multiply or divide mantises and add or subtract exponents. This will be demonstrated in Chapter 5. Analyzing the complexity of the algorithms in the previous chapter, we tend toward one of the forms (4.1) recalled here:

$$A = M_{2's} \cdot 2^{E_{2's}} \text{ or } A = M_{SM} \cdot 2^{E_{2's}} \text{ g}$$

Example 4.2: Number -1.5_{DEC} in floating-point format $M_{SM} \cdot 2^{E_{SM}}$ and n = 4, m = 3.

$$-1.5_{DEC} = -0.75 * 2^1 _{DEC} = 1.110_{SM} * 2^{001}_{SM}.$$

 INTERESTING FACTS!

Alternatives to the floating-point format are the floating slash and signed logarithm presented in [Koren 2002, Matula 1985 and Swartzlander 1975]. However, they have not been widely accepted and are not commonly occurring in everyday practice. On the other hand, the work [Ruszkowski 1983] presents the use of a format with sign-magnitude features for BCD numbers with a floating comma designed for calculators.

4.2 IEEE 754 STANDARD

Most of today's processors have in their structure an additional unit, the so-called FPU performing arithmetic operations on numbers in floating-point format according to IEEE 754:1985 standard. An example is very popular processors from Intel, AMD or ARM64 processor architecture, which in some variants even contain several such units. The lack of an FPU on board the processor, sometimes called an arithmetic coprocessor, does not exclude the possibility of using a floating-point format. Modern compilers of high-level languages have the ability to create machine code for ALU, emulating the lack of FPU, thanks to a dedicated mathematical library of functions, e.g. 'math.c' created for C language. However, the compilation of even a simple program operating on real variables (in floating-point format) results in the generation of extensive and usually unreadable code. The reason for preparing and releasing IEEE 754 was the lack of compatibility between different machines and languages. Such early computers with own real-ization of floating format were, e.g.:

- ODRA 1003/1013 – 39-bit format with mantissa and exponent as 2's numbers (Poland, 1962 year),
- DEC VAX – 4 formats 32/64/64/128 bits, mantissa as SM number, biased exponent (USA, 1977 year),
- IBM 360/370 – 2 formats 32/64 bits, radix 16 instead of 2, mantissa as SM number, biased exponent (USA, 1964 year).

A key milestone was the release three years earlier by Intel of the 8087 chip as a supporting coprocessor of the popular CPU chip Intel 8086. In a way, the IEEE 875:1985 standard is a carry-over of many of the solutions from that chip specification, without reference to implementation details. This leaves technological freedom to other processor manufacturers.

Standard states that number is stored in memory or registers using 3-bit fields: sign, mantissa and exponent. Let us introduce the following rules:

- sign bit S,
- significant M,
- exponent E.

The scientific format is used and value of number can be obtained with the following formula (4.3):

$$A = (-1)^s \cdot M \cdot 2^{E-\text{bias}} \tag{4.3}$$

The sign field needs no comment. Recall only that in the sign-magnitude format S = 1 is for negative number and S = 0 for a non-negative number. The mantissa is re-presented in fixed-point format with a single bit in the integer part and many bits in fraction. Its value is determined by the formula (4.4):

$$M = m_0 + m_1 \cdot 2^{-1} + m_2 \cdot 2^{-2} + \dots \quad m_k \cdot 2^{-k} = m_0 + \sum_{i=1}^{k} m_i \cdot 2^{-i} \tag{4.4}$$

hence the range is $1 \le M < 2$ and $m_0 = 1$ is for normalized number ($m_0 = 0$ is for denormalized). The standard defines four floating-point formats, distinguishing between basic and extended, single or double precision, resulting in four combinations – the first column of Table 4.1.

The standard does not require implementation of the extended format, although it strongly recommends its use to increase the precision of expressing numbers. One of the reasons for inventing the extended format was the need to ensure that computers are comparable (or preferably

Table 4.1 Properties of Normalized Numbers in Floating-point Format According to IEEE 754

Precision	Word length [bits]	The sign 'S' [bits]	The significand 'M'		Exponent "E"	
			Length [bits]	The accuracy for decimal format [significant digits]	Length [bits]	range
Single	32	1	23	7	8	$2^{\pm127} \approx 10^{\pm38}$
Single extended	≥ 43	1	≥ 31	≥ 10	≥ 11	$\geq 2^{\pm1023} \approx 10^{\pm308}$
Double	64	1	52	16	11	$2^{\pm1023} \approx 10^{\pm308}$
Double extended	≥ 79	1	≥ 63	≥ 19	≥ 15	$\geq 2^{\pm16383} \approx 10^{\pm4932}$
Double extended*	80	1	63 + 1	19	15	$2^{\pm16383} \approx 10^{\pm4932}$

higher) with precision with universal calculators. The typical calculator displays a number to ten decimal places, but internally performs operations to 13 digits. Comparing the data in the last column of Table 4.1, it can be seen that a computer working with single-precision numbers has lower calculation accuracy than calculator! The IEEE standard specifies only the minimum number of bits of extended formats, leaving the implementation details to processor and software tool manufacturers (the second column of Table 4.1). Most computer systems are compliant with Intel's implementation, widespread with the 8087 coprocessor. Modern Intel Core-class processors and their clones include such a coprocessor (or several) in their structure, called the FPU. Intel uses the 80-bit format for double extended precision. Wherever we talk about double extended precision conforming to Intel's specification, a '*' will appear to distinguish it from the strict guidelines of the standard. This designation appears, e.g., in Table 4.1. Because the highest bit of the mantissa has always value of 1, the developers of the 8087 coprocessor decided to generally omit it in the bit word. It is only given in the double extended representation, used typically in internal calculations of FPU. In the most cases, numbers are passed to FPU as a single or double precision constant or variables declared in high-level programming language and software.

In addition to Intel's proposal, there are other solutions, also meeting the conditions specified in the standard, but differing in, for example, the number of mantissa bits. For example, HP 700/800 series machines use the following format, called 'quad precision', i.e.: 1 bit – sign bit, 15 bits – exponent, 112 bits – mantissa. For 128-bit number, the four 32-bit width memory locations are needed to store a number value.

Let's go back to the IEEE 754 standard: while the number of exponent bits affects the range of a number, the precision is determined by the

Single precision

	S	E		M	
	31	30	23	22	0

Double precision

	S	E		M	
	63	62	52	51	0

Double extended*

S	E		m_0	M	
79	78	64	63	62	0

Figure 4.2 Format of bit fields according to the IEEE 754.

amount of mantissa bits. For example, for double precision, the mantissa is stored using 53 (52 plus 1 hidden) bits, allowing 2^{53}, or approximately 10^{16} combination of values. Precision, in terms of decimal significant digits, can also be determined in another way, directly from the properties of the number system. Well, x digits can be used to express p^x different values, where p is the base of the system. How many bits are needed to encode one decimal digit? To get the answer, solve the equation $10^1 = 2^x$ with respect to x. If we logarithm it both ways with base 2, we get $x = \log_2(10)$ bits.

So for double precision from the ratio, we get:

- 1 decimal digit – $\log_2(10)$ bits
- y decimal digits – 53 bits

hence y decimal digits $= 53/\log_2(10) = 53/\log_{10}(2) = 15.96 = 16$ digits. Analogous calculations can be done for the other defined formats.

The format of the IEEE 754-bit fields is given in Figure 4.2, and from the table in Appendix B, the number ranges and names of numeric variables can be read, including floating-point, as used in popular high-level languages.

The last field of the number is the exponent stored in the bias format. The use of such a notation, instead of the commonly used sign-magnitude or 2's complement, is related to the necessity of reserving two combinations of exponent bits for special values. The advantage of the bias notation (like for BIN format also) is its monotonicity, which unfortunately is not feature of SM and 2's (Table 4.2). Between the 00h and FFh values, there is a monotonic region of numerical values (for the bias code), which allows to exclude the 00h and FFh boundary values from the allowed numerical range and reserve them for the mentioned special values.

Unfortunately, the disadvantage of this notation is, in the general case, a greater degree of complication of arithmetic operations than 2's or SM. The bias is precision dependent and equal to:

Table 4.2 Comparison of the Variability of Numbers in the Range 00h...FFh for Different Notations

Value	DEC	Biased BIN*	2's complement	Sign-magnitude
Highest	+128	FFh	–	–
	+127	FEh	7Fh	7Fh
	+1	80h	01h	01h
↑	0	7Fh	00h	00h
				80h
	–1	7Eh	FFh	81h
	–127	00h	81h	FFh
Lowest	–128	–	80h	–

* bias 127_{DEC} = $7F_{HEX}$.

- $2^7 - 1 = 127$ for single precision,
- $2^{10} - 1 = 1023$ for double precision,
- $2^{14} - 1 = 16383$ for double extended precision*.

The numbers with exponent with all bits are not zeros or ones are called normalized values. In addition to these, the IEEE 754 standard defines special cases, among which we can distinguish the 0 and ∞ and others, which are given in detail in Table 4.3.

The infinity occurs when the result of an operation exceeds the largest normalized value or an attempt to divide not zero number by zero has occurred, including ∞/0 = ∞. A special combination of bits is reserved for zero because it is impossible to express its value within the accepted normalized number format. Zeroing the fractional part of the mantissa is not sufficient because a bit equal to 1 is assumed in its integer part. The standard also

Table 4.3 Special Values

Sign	Exponent	Mantissa		Value
		m_0	$m_1... m_k$	
1	1 ... 1	1*	0 ... 0	−∞
0	1 ... 1	1*	0 ... 0	+∞
?	1 ... 1	1*	≠0	QNaN (ang. *quiet not a number*)
?	1 ... 1	1*	≠0	SNaN (ang. *signaling not a number*)
1	0 ... 0	0*	0 ... 0	−0
0	0 ... 0	0*	0 ... 0	+0
1	0 ... 0	0	≠0	− denormalized number
0	0 ... 0	0	≠0	+ denormalized number

* Accepted and widely used m_0 values by Intel, among others, although IEEE 754 does not explicitly specify them.

defines two values that have no numerical interpretation, the so-called QNaN and SNaN. Although it does not specify how to encode both types of 'Not a Number', it is generally accepted to distinguish them by the highest bit of the fractional part of the mantissa, i.e.: 0 – SNaN, 1 – QNaN. 'Not a Number' finds many applications. The standard leaves the way they are handled to the processor or compiler manufacturer, requiring only that the silent NaN pass through most arithmetic operations and conversions between formats, thus allowing retrospective analysis of program running and detection of the moment when an undefined value of variable appears. The occurrence of SNaN usually means an invalid value and generates an exception handling. An example of SNaN usage is variable initialization. If program does not assign a value to a variable, it will contain SNaN, which will cause the computation to abort and bug reporting.

The purpose of introducing the concept of 'denormalized numbers' in the standard IEEE 754 requires some comment. These denormalized numbers are also numbers in floating-point format, filling the gap between the smallest normalized value and zero (on both sides of zero). They are encoded by zeros in the exponent field and a non-zero mantissa value. Unlike normalized numbers, the highest, i.e. m_0, bit of the mantissa is assumed to be zero. By introducing denormalized values, you get a gradual transition from normalized numbers to zero. Unfortunately, the closer to zero the number is, the less accurate it is. A denormalized number appears when there is an underflow, i.e., the result of the operation is non-zero and can still be written by denormalizing the mantissa. We encounter such a situation when comparing or subtracting two numbers with close values. If $X \approx Y$, then we should get $X - Y \approx 0$. Not accepting denormalized numbers, we should expect an incorrect result $X - Y = 0$ due to the need to round to zero the result of subtraction with a value smaller than the smallest allowed normalized value. The result of comparing or subtracting two values occurs quite often in algorithms, e.g., in a pair with a conditional jump as a realization of a typical 'if condition then go to' instruction. As a result of rounding the result to zero, the program will run differently than assumed by programmer. This risk is reason of introducing denormalized numbers.

The standard defines five types of exceptions that must be detected and signaled:

- invalid operation, such as:
  - an operation whose argument is SNaN,
  - addition or subtraction of type $(+\infty) + (-\infty)$,
  - multiplication or division: $0*\infty$, $0/0$, ∞/∞,
  - the remainder of dividing x/y when $x = \infty$ or $y = 0$,
  - square root of x for $x < 0$,
  - inability to convert a floating-point number to integer or BCD,
  - inability to compare two data when at least one of them is not a valid number,

- division by zero,
- overflow,
- underflow,
- inaccurate result.

The imprecise result exception occurs when the result of an operation cannot, without loss of precision, be accurately expressed in an accepted format, e.g. the mantissa of $1/3_{DEC}$. A rounded result is usually acceptable in most applications, such as science, where double precision is the standard to ensure accuracy by a large margin. For this reason, this exception as being masked is not handled. Instead, it is supported by applications that are required rigorously processing on accurate (unrounded) numbers. The solution can be, e.g., expressing number like $1/3_{DEC}$ as ratio of two exact numbers 1_{DEC} and 3_{DEC}. It is so called rational format.

 INTERESTING FACTS!

The floating-point units (FPUs) built into Intel processors support an additional type of exception, the denormalized operand exception. Invalid operation, divide-by-zero, and denormalized operand exceptions are pre-computation exceptions and are post-computation exceptions.

By looking at Table 4.1 and using the formula for expressing the value of a number in IEEE 754 format, it is easy to determine the lowest and highest value of a number. Let's look for them for each precision individually.

4.2.1 Single Precision

The combination 11111111, reserved for the special value inf, cannot be used, so the largest value of E is 11111110_{BIN}, hence E – bias = 254 – 127 = 127. The highest mantissa consists of 24 ones (23 bits plus 1 hidden bit in the integer part), hence:

$$M = 1.11111111111111111111111_{SM} = (2^{24} - 1)/2^{23}$$

We calculate the value of the highest normalized number:

$$\begin{aligned}
A_{max\_norm\_single} &= M{\cdot}2^{E-bias} = (2^{24} - 1)/2^{23}{\cdot}2^{127} = (2^{24} - 1){\cdot}2^{104} \\
&= 3.4028234663852885981170418348452{\cdot}10^{38} \\
&\approx 3.4{\cdot}10^{38}
\end{aligned}$$

4.2.2 Double Precision

The highest value of E is 11111111110_{BIN}, hence E − bias = 2046 − 1023 = 1023.

$$M = 1.1111111111111111111111111111111111111111111111111111_{SM}$$
$$= (2^{53} - 1)/2^{52}$$

The value of the largest normalized number is:

$$A_{max\_norm\_double} = M \cdot 2^{E-bias} = (2^{53} - 1)/2^{52} \cdot 2^{1023} = (2^{53} - 1) \cdot 2^{971}$$
$$= 1.797693134862315708145274237317 \cdot 10^{308}$$
$$\approx 1.8 \cdot 10308$$

4.2.3 Double Extended Precision*

The highest value of E is 111111111111110_{BIN}, hence E−bias = 32766 − 16383 = 16383.

$$M = 1.1111111111111111111111111111111111111111111111111111111$$
$$11111111111_{SM}$$
$$= (2^{64} - 1)/2^{63}$$

The value of the highest normalized number is:

$$A_{max\_norm\_double\_extended} = M \cdot 2^{E-bias}$$
$$= (264 - 1)/263 \cdot 216383 = (2^{64} - 1) \cdot 2^{16320}$$
$$\approx 1.18973149535723176502126385303 \cdot 10^{4932}$$

We can repeat similar calculations for denormalized numbers determining the lowest value different from zero.

4.2.4 Single Precision

Although the combination 0...0 is reserved to distinguish the denormalized numbers, it is not used to calculate its value. Therefore, the smallest permitted value of the E field is 1, hence E−bias = 1 − 127 = −126. The mantissa has a zero in integer bit position and 1 in the lowest bit position, therefore:

$$M = 0.00000000000000000000001_{SM} = 2^{-23}$$

The value of the lowest denormalized number different from zero is:

$$A_{min\_denorm\_single} = 2^{-23} \cdot 2^{-126} = 2^{-149}$$
$$= 1.401298464324817070923729583289 \cdot 10^{-45}$$
$$\approx 1.4 \cdot 10^{-45}$$

4.2.5 Double Precision

$E = 00000000001_{BIN}$, hence $E-bias = 1 - 1023 = -1022$.

$M = 0.0000000000000000000000000000000000000000000000000001_{SM}$
$= 2^{-52}$

The value of the lowest denormalized number different from zero is:

$A_{min\_denorm\_double} = 2^{-52} \cdot 2^{-1022} = 2^{-1074}$
$= 4.940656458412465441765687928 6822 \cdot 10^{-324}$
$\approx 4.9 \cdot 10^{-324}$

4.2.6 Double Extended Precision

$E = 000000000000001_{BIN}$, hence $E-bias = 1 - 16383 = -16382$.

$M = 0.0000000000000000000000000000000000000000000000000000000$
0000000001_{SM}
$= 2^{-63}$

The value of the lowest denormalized number different from zero is:

$A_{min\_denorm\_double\_extended} = 2^{-63} \cdot 2^{-16382} = 2^{-16445}$
$= 3.6451995318824746025284059336194$
$\cdot 10^{-4951} \approx 3.6 \cdot 10^{-4951}$

The lowest and highest values for normalized and denormalized numbers are summarized in Table 4.4.

In 1987, the ANSI committee together with the IEEE organization published a standard designated IEEE 854 and entitled 'The IEEE Standard for Radix-Independent Floating-Point Arithmetic'. Unlike the IEEE 754 standard, it allows any integer to be used as the basis of the system, which in fact legitimizes hardware or software implementation of decimal arithmetic. However, it does not specify the details of the basic and extended formats, imposing only the conditions that must be met by the exponent and mantissa of a floating-point number. Those interested in the details of the standard are referred to the source publication [IEEE 1987]. In response to market needs, IEEE 754 was updated in 2008 [IEEE 2008]. Among other things, the extended single-precision format, which had not found acceptance in programming languages, was cancelled, synonyms for 'single' were introduced as equivalent to 'binary32', 'double' was replaced by 'binary64', 'double extended' was replaced by 'binary128', and 'double extended' was replaced by 'extended'. The biggest change, however, was the introduction of two 16-bit formats to support low-cost 16-bit processors used in, e.g.,

Table 4.4 The Lowest and Highest Positive Number According to IEEE 754 Standard

Single precision	Hexadecimal format	Value
The lowest denormalized number	0000 0001	2^{-149}
The highest denormalized number	007F FFFF	$2^{149} \cdot (2^{23} - 1)$
The lowest normalized number	0080 0000	2^{-126}
The highest normalized number	7F7F FFFF	$2^{104} \cdot (2^{24} - 1)$
Double precision	**Hexadecimal format**	**Value**
The lowest denormalized number	0000 0000 0000 0001	2^{-1074}
The highest denormalized number	000F FFFF FFFF FFFF	$2^{1074} \cdot (2^{52} - 1)$
The lowest normalized number	0010 0000 0000 0000	2^{-1022}
The highest normalized number	7FEF FFFF FFFF FFFF	$2^{971} \cdot (2^{53} - 1)$
Double precision*	**Hexadecimal format**	**Value**
The lowest denormalized number	0000 0000 0000 0000 0001	2^{-16445}
The highest denormalized number	0000 7FFF FFFF FFFF FFFF	$2^{16445} \cdot (2^{63} - 1)$
The lowest normalized number	0001 8000 0000 0000 0000	2^{-16382}
The highest normalized number	7FFE FFFF FFFF FFFF FFFF	$2^{16320} \cdot (2^{64} - 1)$

cash register, parking meter, ticket machine, water and gas consumption meters, etc. The double extended precision format (now called extended) has had its bit count increased to 128. The precision and numeric ranges of these new formats are also included in Table B.2 of Appendix B. It is worth mentioning that not all manufacturers of FPUs and programming tools have decided to fully implement the recommendations of the standard in 2008 version. That is main reason why the original version of IEEE 754 was presented in this book. Thus, the 16-bit format is present, e.g. in MATLAB, GIMP packages, Direct3D, D3DX, OpenGL, Cg (NVIDIA & Microsoft) libraries, and OpenEXR and JPEG XR graphic file formats. The only operation on this format in the FPU of Intel Core processors is the conversion to and from 32-bit format. Examples of hardware implementation of operations on 128-bit format are the following families of CPUs: Intel Core, IBM Power P9 and Fujitsu SPARC V8/9. The new version of standard specifies additional operations that are recommended for all supported arithmetic formats. These operations are given as function names, but in a particular programming environment they may be represented by operators or functions whose names may differ. These include, among others:

- $\exp(x)$, 2^x, 10^x, $\ln(x)$, $\log_2(x)$, $\log_{10}(x)$,
- $\operatorname{sqrt}(x^2 + y^2)$, $1/\operatorname{sqrt}(x)$, $x^{1/n}$,
- $\sin(x)$, $\cos(x)$, $\tan(x)$, $\operatorname{asin}(x)$, $\operatorname{acos}(x)$, $\operatorname{atan}(x)$,
- $\sinh(x)$, $\cosh(x)$, $\tanh(x)$, $\operatorname{asinh}(x)$, $\operatorname{acosh}(x)$, $\operatorname{atanh}(x)$.

The reader will find more information in the reference publication [IEEE 2008].

4.3 FPU AS A SPECIALIZED ARITHMETIC UNIT

An FPU is like a 'younger brother' of classical ALU but is a much powerful. It extends the processor arithmetic abilities to work on high precision a wide range real numbers. It rather occurs as resource of families of strong processor, e.g. Intel, AMD, NXP ColdFire or ARM architectures. The details of an internal architecture differ but from the user point of view they provide similar functionalities thanks to conforming to requirements of IEEE 754 standard (at least to the most of them). For example, Intel FPU, called x87 FPU, consists of eight 80-bit data registers and special-purpose registers for managing the rounding modes, exceptions and many others. Values are stored in these registers in the double extended precision format – look for Table 4.1 and comments on '*'. When floating-point, integer or packed BCD values are loaded from memory into any of FPU data registers, the values are automatically converted into double extended precision floating-point format or not if they are already in that format. When computation results are sending back into memory from any of the x87 FPU registers, the results can be left in the double extended precision floating-point format or converted back into a shorter floating-point format, an integer format, or the packed BCD integer format. The x87 FPU instructions treat the eight x87 FPU data registers as a classical register stack. Addressing of the data registers is relative to the register on the top of the stack. If a load operation is performed when top of a stack is at R0 register, the register wraparound occurs and the new localization of stack top is assigned to R7.

The ARM architectures support floating-point data types and arithmetic with some restrictions depending on specific core version. For example, the Armv8 architecture supports both single and double precision data types. It also supports the 16-bit half-precision floating-point data type for data storage, by supporting conversions between single-precision and half-precision data types and double-precision and half-precision data types. Another example is the ARM Cortex-M4 core. Its FPU fully supports only single-precision add, subtract, multiply, divide, multiply and accumulate, and square root operations. It also provides conversions between fixed-point and floating-point data formats, and floating-point constant instructions. The FPU provides an extension register file containing 32 single-precision registers. These can be viewed as:

- sixteen 64-bit double-word registers, D0-D15,
- thirty-two 32-bit single-word registers, S0-S31,
- or combination of registers from the above views.

Specific options beyond the standard are 'flush-to-zero' and 'default NaN' modes. In 'flush-to-zero' mode, the FPU treats all denormalized input operands of arithmetic operations as zeros. For the 'default NaN' mode the result of any arithmetic data processing operation that involves an input NaN, or that generates a NaN result, returns the default NaN. The default NaN is a qNaN with an all-zero of mantissa fraction. When not in default NaN mode, the operations with NaN input values preserve the NaN, or one of the NaN values, if more the one input operand is a NaN, as the result [Hohl 2015].

4.4 CONVERSION TO ANOTHER RADIX

Conversion of a floating-point number A with base p to base s involves finding the value of the mantissa M_s and the exponent E_s according to the formulas (4.5):

$$C_s = 1 + \frac{\ln |M_p| + C_p \cdot \ln(p)}{\ln(s)}$$

$$M_s = M_p \cdot \exp(C_p \cdot \ln(p) - C_s \cdot \ln(s)) \tag{4.5}$$

for those the below equality is satisfied (4.6):

$$A = M_p \cdot p^{C_p} = M_S \cdot s^{C_S} \tag{4.6}$$

The derivation of the given formulas will be presented here. Let's an A number be as below (4.7):

$$A = M_S \cdot s^{C_S} \tag{4.7}$$

assuming that sign is part of mantissa.

Logarithmizing both sides at the base s of the above relation (considering the domain of the logarithm), and then adding a constant 1 to both sides, we get (4.8):

$$1 + \log_s(|A|) = \log_s(|M_S|) + \underbrace{\log_s(s)}_{1} + C_S = \log_s(s \cdot |M_S|) + C_S \tag{4.8}$$

Because of observation that $\log_s(s \cdot |M_S|) < C_S$, we obtain in simplified form (4.9):

$$C_S \approx 1 + \log_s(A) = 1 + \frac{\ln(|A|)}{\ln(s)} \tag{4.9}$$

Knowing that $A = M_p \cdot p^{C_p}$ and applying the rule of changing the logarithm base, we have (4.10):

$$C_S \approx 1 + \log_s(A) = 1 + \frac{\ln(|A|)}{\ln(s)} = 1 + \left[\frac{\ln(|M_p|) + C_p \cdot \ln(p)}{\ln(s)}\right] \qquad (4.10)$$

The identity $x = \exp(\ln(x))$ for $x > 0$ yields the final mantissa formula with base s as shown by (4.11):

$$M_S = M_p \cdot \frac{p^{C_p}}{s^{C_S}} = M_p \cdot \exp(C_p \cdot \ln(p) - C_s \cdot \ln(s)) \qquad (4.11)$$

In general, the accepted format for floating-point numbers requires C_S to be an integer, so it is sometimes necessary to round the value obtained from the calculation. It follows as a practical matter to first determine the exact value of the characteristic C_s, then round it, and finally determine the mantissa M_S.

Example 4.3: The floating-point number $\underline{0}.1101_{SM} \cdot 2^{\underline{0}11}{}_{SM} = 6.5_{DEC}$ with base 2 and the same number with base 10:

$M_2 = \underline{0},110_{SM} = +13/16_{DEC}$ and $C_2 = \underline{0}11_{SM} = +3_{DEC}$.

$$C_{10} = 1 + \frac{\ln|M_2| + C_2 \cdot \ln 2}{\ln 10} = 1 + \frac{\ln|+0.8125| + (+3) \cdot \ln 2}{\ln 10} = 1.81 \approx 2$$

$$M_{10} = M_2 \cdot \exp(C_2 \cdot \ln 2 - C_{10} \cdot \ln 10) = +0.8125 \cdot \exp(3 \cdot \ln 2 - 2 \cdot \ln 10) = +0.065$$

Finally, we have $+0.065 \cdot 10^{+2}$. Checking: $+0.065 \cdot 10^{+2} = +6.5_{DEC}$

Example 4.4: The floating-point number $-3.72 \cdot 10^{-2}$ with base 10 and the same number with base 2 in SM format:

$M_{10} = -3.72_{DEC}$ and $C_{10} = -2_{DEC}$.

$$C_2 = 1 + \frac{\ln|M_{10}| + C_{10} \cdot \ln 10}{\ln 2} = 1 + \frac{\ln|-3.72| + (-2) \cdot \ln 10}{\ln 2} = -3.747 \approx -4$$
$$= 1100_{SM}$$

$$M_2 = M_{10} \cdot \exp(C_{10} \cdot \ln 10 - C_2 \cdot \ln 2) = -3.72 \cdot \exp(-2 \cdot \ln 10 - (-4) \cdot \ln 2)$$
$$= -0.595_{DEC}$$

0.595·2
$\underline{1}$.19·2
$\underline{0}$.38·2
$\underline{0}$.76·2
$\underline{1}$.52·2
$\underline{1}$.04·2

...

$0.595_{DEC} \rightarrow \approx 0.10011_{BIN}$

Finally we have $\underline{1}.10011_{SM} \cdot 2^{\underline{1}100}{}_{SM}$. Checking: $-0.595 \cdot 2^{-4} = -0.0372_{DEC}$

Exercise 4.1: Convert the floating-point number $-5.28 \cdot 10^{-3}$ with base 10 to the base of 2 in SM format.

Exercise 4.2: Convert the floating-point number $\underline{0}.0101_{SM} \cdot 2^{\underline{0}10}{}_{SM} = +1.25_{DEC}$ with base 2 to the base of 10.

Chapter 5

Basic Arithmetic Operations on Floating-point Numbers

5.1 ADDITION

The purpose of the following discussion is to show how to perform the operation $Z = X + Y$, where (5.1):

$$M_Z \cdot p^{E_Z} = M_X \cdot p^{E_X} + M_Y \cdot p^{E_Y}$$

$$M_Z = M'_X + M'_Y \quad E_Z = \max(E_X, E_Y) \tag{5.1}$$

and (5.2)

$$\begin{array}{lll} M'_X = M_X, & M'_Y = M_Y & \text{for } E_X = E_Y \\ M'_X = M_X, & M'_Y = M_Y \cdot p^{-|C_X - C_Y|} & \text{for } E_X > E_Y \\ M'_X = M_X \cdot p^{-|C_X - C_Y|}, & M'_Y = M_Y & \text{for } E_X < E_Y \end{array} \tag{5.2}$$

Addition of two mantises should be done according to the rules specified in Chapter 3. Mantises in 2's format are added according to the formula $M_Z = M'_X + M'_Y$. Adding the mantises in SM format requires applying the rules given in Table 3.1. Sometimes the result of the addition cannot be stored on the assumed number of bits, so it should be normalized according to the following rules:

- if $|M_Z| \geq 1$ then $M_{Z\_norm} = M_Z \cdot p^{-1}$ $E_{Znorm} = E_Z + 1$,
- if $|M_Z| < p^{-i}$ then $M_{Z\_norm} = M_Z \cdot p^i$ $E_{Znorm} = E_Z - i$, and i – number of zeros after dot point, e.g. $i = 3$ for 0.0001.

Example 5.1: Addition of two floating-point SM numbers:

$$X = \underline{0}.1100 \cdot 2^{111}{}_{SM} = +\frac{12}{16} \cdot 2^{-3} = +0.09375_{DEC}, \quad Y = \underline{0}.0001 \cdot 2^{010} = +\frac{1}{16} \cdot 2^{+2}$$

$$= +0.25_{DEC}$$

DOI: 10.1201/9781003363286-5

For better readiness, we will sometimes replace the sign bit values 0/1 with +/−.

We have $p = 2$, $E_X = -11_{SM}$, $M_X = +0.1100_{SM}$, $E_Y = +10_{SM}$, $M_Y = +0.0001_{SM}$.

$$|E_X - E_Y| = |-3 - (+2)| = |-5| = +5, \quad E_X < E_Y \quad E_Z = max(-3, +2) = +2$$

$$M'_X = M_X \cdot 2^{-5} = \underline{0}.0000011_{SM}, \quad M'_Y = M_Y = \underline{0}.0001000_{SM}$$

Because of X and Y are positive (see addition rules for SM numbers in Table 3.1):

$$M_Z = M'_X + M'_Y = +(|M'_X| + |M'_Y|) = \underline{0}.0001011_{SM}$$

We have $|M_Z| < 2^{-3}$ therefore the normalization is needed:

$$M_{Z\_norm} = M_Z \cdot 2^3 = \underline{0}.1011_{SM}, \quad E_{Znorm} = E_Z - 3 = +2 - 3 = -1 = \underline{1}0_{SM}$$

Finally, $Z = \underline{0}.1011 \cdot 2^{\underline{1}01}{}_{SM} = +0.34375_{DEC}$.

Exercise 5.1: Perform addition of two floating-point SM numbers:

$$X = \underline{1}.10111 \cdot 2^{\underline{0}10}{}_{SM} = -\frac{11}{16} \cdot 2^{+2} = -2.75_{DEC}, \quad Y = \underline{0}.11111 \cdot 2^{\underline{1}01}{}_{SM} = +\frac{15}{16} \cdot 2^{-1}$$
$$= +0.46875_{DEC}$$

5.2 SUBTRACTION

The purpose of the following discussion is to show how to perform the operation $Z = X - Y$, where (5.3)

$$M_Z = M'_X - M'_Y \quad E_Z = max(E_X, \ E_Y) \tag{5.3}$$

and (5.4)

$$
\begin{array}{llll}
M'_X = M_X, & M'_Y = M_Y & \text{dla } E_X = E_Y & \\
M'_X = M_X, & M'_Y = M_Y \cdot p^{-|E_X - E_Y|} & \text{dla } E_X > E_Y & (5.4) \\
M'_X = M_X \cdot p^{-|E_X - E_Y|} & M'_Y = M_Y & \text{dla } E_X < E_Y &
\end{array}
$$

Subtraction of two mantises should be done according to the rules specified in Chapter 3. Mantises in 2's format are subtracted according to the formula $M_Z = M'_X - M'_Y$. Subtracting the mantises in SM format requires applying the rules given in Table 3.2. Normalization of the result follows the rules specified in the previous chapter.

Example 5.2: Subtraction of two floating-point SM numbers:

$$X = 1.100 \cdot 2^{001}_{SM} = -\frac{1}{2} \cdot 2^{+1} = -1_{DEC},$$

$$Y = 0.111 \cdot 2^{010}_{SM} = +\frac{7}{8} \cdot 2^{+2} = +3.5_{DEC}.$$

$$p = 2, \quad E_X = +01_{SM}, \quad M_X = -0.100_{SM},$$
$$E_Y = +10_{SM}, \quad M_Y = +0.111_{SM}$$

$$|E_X - E_Y| = |+1 - (+2)| = |-1| = +1,$$
$$E_X < E_Y \quad E_Z = \max(+1, +2) = +2$$

$$M'_X = M_X \cdot 2^{-1} = -0.010_{SM}, \quad M'_Y = M_Y$$

Because of X is negative and Y positive (see subtraction rules for SM numbers in Table 3.2):

$$M_Z = M'_X - M'_Y = -(|M'_X| + |M'_Y|) = -1.001_{SM}$$

We have $|M_Z| \geq 1$, therefore the normalization is needed:

$$M_{Z\_norm} = M_Z \cdot 2^{-1} = -0.100_{SM} = 1.100_{SM}, \quad E_{Znorm} = E_Z + 1 = +2 + 1$$
$$= 011_{SM}$$

If an equal number of bits of the result Z and the arguments X, Y are assumed, then the lowest bit of the mantissa must be discarded. As a result, the result will be approximated. Finally, $Z \approx 1.100 \cdot 2^{011}_{SM} = -4_{DEC}$ but we expected -4.5_{DEC}.

Exercise 5.2: Perform subtraction of two floating-point SM numbers:

$$X = 1.101 \cdot 2^{000}_{SM} = -\frac{5}{8}_{DEC}, \quad Y = 1.110 \cdot 2^{010}_{SM} = -\frac{6}{8} \cdot 2^{+2} = -3_{DEC}$$

5.3 MULTIPLICATION

The purpose of the following discussion is to show how to perform the operation $Z = X \cdot Y$ (5.5):

$$M_Z \cdot p^{E_Z} = \left(M_X \cdot p^{E_X} \right) \cdot \left(M_Y \cdot p^{E_Y} \right)$$

$$M_Z = M_X \cdot M_Y \quad E_Z = E_X + E_Y \tag{5.5}$$

Normalization of the result of mantissa multiplication follows the rules specified for addition of floating-point numbers. The multiplication does not require a preliminary denormalization of the mantissa, which simplifies the execution of the operation.

Example 5.3: Multiplication of two floating-point SM numbers, where

$$X = 1,0110 \cdot 2^{101}_{SM} = -\frac{3}{8} \cdot 2^{-1} = -\frac{3}{16}_{DEC},$$

$$Y = 0,1010 \cdot 2^{011}_{SM} = +\frac{5}{8} \cdot 2^{+3} = +5_{DEC}$$

Data:

$$p = 2, \quad E_X = -0_{SM}, \quad M_X = -0.0110_{SM},$$
$$E_Y = +1_{SM}, \quad M_Y = +0.1010_{SM}$$

$$E_Z = E_X + E_Y = -1 + (+3) = +2_{DEC}$$

Because X and Y are different signs (see multiplication rules for SM format):

$$M_Z = M_X \cdot M_Y = -|M_X| \cdot |M_Y| = -0.00111100_{SM}$$

We have $|M_Z| < 2^{-2}$, hence the normalization is needed:

$$M_{Z\_norm} = M_Z \cdot 2^2 = -0.1111_{SM} = 1.1111_{SM},$$
$$E_{Znorm} = E_Z - 2 = +2 - 2 = 0 = \underline{0}00_{SM}$$

Finally, $Z = 1.1111 \cdot 2^{000}_{SM} = -\dfrac{15}{16}_{DEC}.$

Exercise 5.3: Perform multiplication of two floating-point numbers, where

$$X = 1.0100 \cdot 2_{SM}^{001} = -\frac{1}{4} \cdot 2^{+1} = -\frac{1}{2}_{DEC},$$

$$Y = 1.0010 \cdot 2_{SM}^{110} = -\frac{1}{8} \cdot 2^{-2} = -\frac{1}{32}_{DEC}$$

5.4 DIVISION

The purpose of the following discussion is to show how to perform the operation $Z = X/Y$ (5.6):

$$M_Z \cdot p^{C_Z} = \left(M_X \cdot p^{C_X}\right)/\left(M_Y \cdot p^{C_Y}\right), \quad C_Z = C'_X - C_Y$$

$$M'_X = M_X, \quad C'_X = C_X \qquad \text{for } |M_X| < |M_Y|$$

$$M'_X = M_X \cdot p^{-j}, \quad C'_X = C_X + j, \quad \text{for } |M_X| \geq |M_Y| \text{ and } \overset{\vee}{j}(|M'_X| < |M_Y|)$$

$$(5.6)$$

The parameter j specifies in practice the number of zeros to be added after the decimal point so that the denormalized mantissa M_X is smaller than the mantissa M_Y in absolute value. The normalization of the result of dividing the mantissa follows the rules defined for adding floating-point numbers.

Example 5.4: Division of two floating-point SM numbers, where

$$X = \underline{0}.1010 \cdot 2_{SM}^{011} = +\frac{5}{8} \cdot 2^{+3} = +5_{DEC},$$

$$Y = 1.0110 \cdot 2_{SM}^{101} = -\frac{3}{8} \cdot 2^{-1} = -\frac{3}{16}_{DEC}$$

$$p = 2, \quad C_X = +11_{SM}, \quad M_X = +0.1010_{SM},$$

$$C_Y = -01_{SM}, \quad M_Y = -0.0110_{SM}$$

Because of $|M_x| > |M_Y|$, we have to firstly make a denormalization of X number, hence

$$\text{for } j = 1 \quad |M'_X| = |M_X \cdot 2^{-j}| \leq |M_Y|$$

and $M'_X = M_X \cdot 2^{-1} = +0.0101_{SM}, \quad C'_X = C_X + 1 = +100_{SM}, \quad C_Z = C'_X - C_Y = +4$

$$- (-1) = +5_{DEC}$$

Because X and Y are of equal signs (see rules for division in SM format):

$$M_Z = M'_X/M_Y = |M'_X|/|M_Y| = -0.11010|_{SM} = \underline{1}.11010|_{SM}$$

In this case, there is no need to normalize the result. If an equal number of bits are assumed for the result Z and the arguments X and Y then the two lowest bits of the mantissa must be discarded. As a result, the value of the mantissa will be approximated. Furthermore, to express a feature of value +5 in SM format, 4 bits are needed and the assumed number is 3, so truncation error of 010_{SM}-$0101_{SM} = 111_{SM}$ would occur. The solution to the problem is to take 4 bits to express the exponents. Finally, $Z \approx \underline{1}.1101 \cdot 2^{0101}{}_{SM} \approx -26.66_{DEC}$. A discussion of the effect of finite precision on the accuracy of the result of arithmetic operations is undertaken in Chapter 6.

Exercise 5.4: Perform division of two floating-point SM numbers, where

$$X = \underline{0}.1101 \cdot 2^{010}_{SM} = +\frac{13}{16} \cdot 2^{+2} = +\frac{13}{4} = +3.25_{DEC}$$

$$Y = \underline{1}.1110 \cdot 2^{\underline{1}10}_{SM} = -\frac{14}{16} \cdot 2^{-2} = -\frac{14}{64}{}_{DEC}$$

5.5 IMPLEMENTATIONS IN ASSEMBLY LANGUAGE

The software implementation of operation on floating-point numbers in the 8051 microcontroller according to the IEEE 754 standard is quite complicated. It requires performing operations on multi-byte numbers, where the mantissa is in SM format and the exponent in biased format. To illustrate the complexity of the problem, we will show the implementation for numbers using simplified format. We present subroutines implementing four basic arithmetic operations for 2-byte numbers, where 1 byte for the mantissa and the other for the exponent are reserved. Comparing the complexity of arithmetic operations for numbers in different formats (see Chapter 3), we propose to express the mantissa and the exponent as SM numbers, i.e., each number will be seen as $A = M_{SM} \cdot 2^{C_{SM}}$, where (5.7):

$$M_{SM} = (-1)^{\tilde{a}_0} \cdot (a_{-1} \cdot 2^{-1} + ... + a_{-(n-1)} \cdot 2^{-(n-1)}) = (-1)^{\tilde{a}_0} \cdot \sum_{i=-1}^{-(n-1)} a_i \cdot 2^i$$

$$C_{SM} = (-1)^{a_{m-1}} \cdot (a_{m-2} \cdot 2^{m-2} + ... + a_1 \cdot 2 + a_0) = (-1)^{a_{m-1}} \cdot \sum_{i=0}^{m-2} a_i \cdot 2^i \quad (5.7)$$

Since a dimension of mantissa and exponent are 1 byte, hence n = 8 and m = 8. The lowest and highest mantissa values are:

- the lowest negative: $\underline{1}.1111111_{SM} = -127/128$,
- the highest negative: $\underline{1}.0000000_{SM} = -0$,
- the lowest positive: $\underline{0}.0000000_{SM} = +0$,
- the highest positive: $\underline{0}.1111111_{SM} = +127/128$.

Let's list the lowest and highest values of the exponent:

- the lowest negative: $\underline{1}1111111_{SM} = -127$,
- the highest negative: $\underline{1}0000000_{SM} = -0$,
- the lowest positive: $\underline{0}0000000_{SM} = +0$,
- the highest positive: $\underline{0}1111111_{SM} = +127$.

Minimal and maximal values for the assumed convention are:

$$0.1111111 \cdot 2_{SM}^{01111111} \approx +10_{DEC}^{+38}$$
$$1.1111111 \cdot 2_{SM}^{01111111} \approx -10_{DEC}^{-38}$$

with resolution of mantissa 1/128 = 0.0078125 less than 2.5 decimal digits.

According to the rules discussed in the previous chapters, the execution of the operations consists in performing an initial denormalization of the mantises, except for multiplication, followed by addition, subtraction, multiplication or division of the mantises, and multiplication of the exponents in the case of multiplication, or denormalization and division of the features in the case of division. Regardless of the type of operation, the result must be reported in normalized form. In the subroutines presented here we will refer to the respective subroutines outlined in Chapter 3. Setting the OV flag will indicate a result out of range.

Implementation in code of the <u>addition</u>:

- input number: R0 – mantissa of first number, R1 – exponent of first number,
- input number: R2 – mantissa of second number, R3 – exponent of second number,
- output number: R0 – mantissa of result, R1 – exponent of result, OV – result out of the range,
- exemplary value: $[+0.625*2^{+3}] + [-0.375*2^{-1}]$.

```
1  ;********************************************************************
2  ;* Addition of floating-point numbers *
3  ;********************************************************************
4  ;mantissa must be normalized, i.e. 1 after sign bit!
```

```
 5 ;an exception is mantissa with value of 0.
 6
 7 0003          c1 EQU 00000011B      ;exponent of first SM number
 8 0050          m1 EQU 01010000B      ;mantissa of first SM number
 9 0081          c2 EQU 10000001B      ;exponent of second SM number
10 00B0          m2 EQU 10110000B      ;mantissa of second SM number
11                ;first number        m1*2^c1=+0.625*2^+3
12                ;second number       m2*2^c2=-0.375*2^-1
13                ;result              my*2^cy=+77/128*2^+3=+4.8125
14                ;another example of numbers
15                ;c1 EQU 10000111B     ;exponent of first SM number
16                ;m1 EQU 11000000B     ;mantissa of first SM number
17                ;c2 EQU 10000100B     ;exponent of second SM number
18                ;m2 EQU 01110000B     ;mantissa of second SM number
19                ;first number        m1*2^c1=-0.5*2^-7
20                ;second number       m2*2^c2=+0.875*2^-4
21                ;result              my*2^cy=+13/16*2^-4=
                                         +0.8125*2^-4
22
23 0000: 78 50    MOV R0,#m1
24 0002: 79 03    MOV R1,#c1
25 0004: 7A B0    MOV R2,#m2
26 0006: 7B 81    MOV R3,#c2
27 0008: 12 00 0D LCALL ADD_FLOATS
28                                     ;result in A
29 000B: 80 FE    STOP: SJMP STOP
30 ;---------------------------------------------------------------------------------
31 000D:          ADD_FLOATS:
32 000D: EB       MOV A,R3
33 000E: 89 F0    MOV B,R1
34 0010: B2 E7    CPL ACC.7
35 0012: 12 00 72 LCALL ADD_SM_FLOATS ;compare the exponents
36 0015: 60 2D    JZ EXP_THE_SAME
37 0017: 92 E7    MOV ACC.7,C
38 0019: 30 E7 17 JNB ACC.7,GREATER
39 001C:          LESS:
40 001C: C2 E7    CLR ACC.7
41 001E: FC       MOV R4,A             ;c1<c2
42 001F: E8       MOV A,R0
43 0020: A2 E7    MOV C,ACC.7
44 0022: 92 D1    MOV PSW.1,C
45 0024: C2 E7    CLR ACC.7
46 0026:          NORM1:
47 0026: C3       CLR C
48 0027: 13       RRC A
```

```
49  0028: DC FC      DJNZ R4,NORM1
50  002A: A2 D1      MOV C,PSW.1
51  002C: 92 E7      MOV ACC.7,C
52  002E: F8         MOV R0,A
53  002F: EB         MOV A,R3
54  0030: C9         XCH A,R1
55  0031: 80 11      SJMP EXP_THE_SAME
56  0033:            GREATER:
57  0033: FC         MOV R4,A            ;c1>c2
58  0034: EA         MOV A,R2
59  0035: A2 E7      MOV C,ACC.7
60  0037: 92 D1      MOV PSW.1,C
61  0039: C2 E7      CLR ACC.7
62  003B:            NORM2:
63  003B: C3         CLR C
64  003C: 13         RRC A
65  003D: DC FC      DJNZ R4,NORM2
66  003F: A2 D1      MOV C,PSW.1
67  0041: 92 E7      MOV ACC.7,C
68  0043: FA         MOV R2,A
69  0044:            EXP_THE_SAME:
70  0044: E8         MOV A,R0
71  0045: 8A F0      MOV B,R2
72  0047: 12 00 72   LCALL ADD_SM_FLOATS
73  004A: 92 D5      MOV PSW.5,C
74  004C: 30 E7 06   JNB ACC.7,NORM3
75  004F: C3         CLR C
76  0050: 03         RR A
77  0051: 7C 01      MOV R4,#1
78  0053: 80 0E      SJMP SKIP
79  0055:            NORM3:
80  0055: 7C 00      MOV R4,#0
81  0057:            RETURN:
82  0057: 20 E6 05   JB ACC.6,SKIP1
83  005A: C3         CLR C
84  005B: 23         RL A
85  005C: 0C         INC R4
86  005D: 80 F8      SJMP RETURN
87  005F:            SKIP1:
88  005F: CC         XCH A,R4
89  0060: B2 E7      CPL ACC.7
90  0062: CC         XCH A,R4
91  0063:            SKIP:
92  0063: A2 D5      MOV C,PSW.5
93  0065: 92 E7      MOV ACC.7,C
```

```
 94  0067: F8         MOV R0,A              ;mantissa of result
 95  0068: E9         MOV A,R1
 96  0069: 8C F0      MOV B,R4
 97  006B: 12 00 72   LCALL ADD_SM_FLOATS
 98  006E: 92 E7      MOV ACC.7,C
 99  0070: F9         MOV R1,A              ;exponent of result
100  0071: 22         RET
101  ;-------------------------------------------------------------------------------
102  0072:            ADD_SM_FLOATS:
103  0072: A2 E7      MOV C,ACC.7
104  0074: 92 D5      MOV PSW.5,C
105  0076: C0 E0      PUSH ACC
106  0078: 65 F0      XRL A,B
107  007A: A2 E7      MOV C,ACC.7
108  007C: 53 F0 7F   ANL B,#01111111B
109  007F: D0 E0      POP ACC
110  0081: 54 7F      ANL A,#01111111B
111  0083: 50 18      JNC signs_the_same
112  0085:            signs_different:
113  0085: B5 F0 02   CJNE A,B,different
114  0088: 80 02      SJMP greater_or_equal
115  008A:            different:
116  008A: 40 07      JC less
117  008C:            greater_or_equal:
118  008C: C3         CLR C
119  008D: 95 F0      SUBB A,B
120  008F: A2 D5      MOV C,PSW.5
121  0091: 80 12      SJMP end
122  0093:            less:
123  0093: C3         CLR C
124  0094: C5 F0      XCH A,B
125  0096: 95 F0      SUBB A,B
126  0098: A2 D5      MOV C,PSW.5
127  009A: B3         CPL C
128  009B: 80 08      SJMP end
129  009D:            signs_the_same:
130  009D: 25 F0      ADD A,B
131  009F: A2 E7      MOV C,ACC.7
132  00A1: 92 D2      MOV OV,C
133  00A3: A2 D5      MOV C,PSW.5
134  00A5:            end:
135  00A5: 22         RET
136  ;--- end of file ---
```

Implementation in code of the <u>subtraction</u>:

- input number: R0 – mantissa of first number, R1 – exponent of first number,
- input number: R2 – mantissa of second number, R3 – exponent of second number,
- output number: R0 – mantissa of result, R1 – exponent of result, OV – result out of the range,
- exemplary value: $[+0.625*2^{+3}] - [-0.375*2^{-1}]$.

```
 1   ;*******************************************************************
 2   ;* Subtraction of floating-point numbers *
 3   ;*******************************************************************
 4   ;mantissa must be normalized, i.e. 1 after sign bit!
 5   ;an exception is mantissa with value of 0.
 6
 7   0003              c1 EQU 00000011B        ;exponent of first SM number
 8   0050              m1 EQU 01010000B        ;mantissa of first SM number
 9   0081              c2 EQU 10000001B        ;exponent of second SM number
10   00B0              m2 EQU 10110000B        ;mantissa of second SM number
11                     ;first number           m1*2^c1=+0.625*2^+3
12                     ;second number          m2*2^c2=-0.375*2^-1
13                     ;result                 my*2^cy=+83/128*2^+3
14                     ;another example of numbers
15                     ;c1 EQU 10000111B       ;exponent of first SM number
16                     ;m1 EQU 11000000B       ;mantissa of first SM number
17                     ;c2 EQU 10000100B       ;exponent of second SM number
18                     ;m2 EQU 01110000B       ;mantissa of second SM number
19                     ;first number           m1*2^c1=-0.5*2^-7
20                     ;second number          m2*2^c2=+0.875*2^-4
21                     ;result                 my*2^cy=-15/16*2^-4
22
23   0000: 78 50       MOV R0,#m1
24   0002: 79 03       MOV R1,#c1
25   0004: 7A B0       MOV R2,#m2
26   0006: 7B 81       MOV R3,#c2
27   0008: 12 00 0D    LCALL SUB_FLOATS
28                                             ;result in A
29   000B: 80 FE       STOP: SJMP STOP
30   ;-------------------------------------------------------------------
31   000D:             SUB_FLOATS:
32   000D: E9          MOV A,R1
33   000E: 8B F0       MOV B,R3
34   0010: 12 00 72    LCALL SUB_SM_FLOATS     ;compare the exponents
35   0013: 60 2D       JZ EXP_THE_SAME
```

```
36    0015: 92 E7      MOV ACC.7,C
37    0017: 30 E7 17   JNB ACC.7,GREATER
38    001A:           LESS:
39    001A: C2 E7      CLR ACC.7
40    001C: FC         MOV R4,A                    ;c1<c2
41    001D: E8         MOV A,R0
42    001E: A2 E7      MOV C,ACC.7
43    0020: 92 D1      MOV PSW.1,C
44    0022: C2 E7      CLR ACC.7
45    0024:           NORM1:
46    0024: C3         CLR C
47    0025: 13         RRC A
48    0026: DC FC      DJNZ R4,NORM1
49    0028: A2 D1      MOV C,PSW.1
50    002A: 92 E7      MOV ACC.7,C
51    002C: F8         MOV R0,A
52    002D: EB         MOV A,R3
53    002E: C9         XCH A,R1
54    002F: 80 11      SJMP EXP_THE_SAME
55    0031:           GREATER:
56    0031: FC         MOV R4,A                    ;c1>c2
57    0032: EA         MOV A,R2
58    0033: A2 E7      MOV C,ACC.7
59    0035: 92 D1      MOV PSW.1,C
60    0037: C2 E7      CLR ACC.7
61    0039:           NORM2:
62    0039: C3         CLR C
63    003A: 13         RRC A
64    003B: DC FC      DJNZ R4,NORM2
65    003D: A2 D1      MOV C,PSW.1
66    003F: 92 E7      MOV ACC.7,C
67    0041: FA         MOV R2,A
68    0042:           EXP_THE_SAME:
69    0042: E8         MOV A,R0
70    0043: 8A F0      MOV B,R2
71    0045: 12 00 72   LCALL SUB_SM_FLOATS
72    0048: 92 D5      MOV PSW.5,C
73    004A: 30 E7 06   JNB ACC.7,NORM3
74    004D: C3         CLR C
75    004E: 03         RR A
76    004F: 7C 01      MOV R4,#1
77    0051: 80 0E      SJMP SKIP
78    0053:           NORM3:
79    0053: 7C 00      MOV R4,#0
80    0055:           RETURN:
```

```
81    0055: 20 E6 05    JB ACC.6,SKIP1
82    0058: C3          CLR C
83    0059: 23          RL A
84    005A: 0C          INC R4
85    005B: 80 F8       SJMP RETURN
86    005D:            SKIP1:
87    005D: CC          XCH A,R4
88    005E: B2 E7       CPL ACC.7
89    0060: CC          XCH A,R4
90    0061:            SKIP:
91    0061: A2 D5       MOV C,PSW.5
92    0063: 92 E7       MOV ACC.7,C
93    0065: F8          MOV R0,A              ;mantissa of result
94    0066: E9          MOV A,R1
95    0067: 8C F0       MOV B,R4
96    0069: B2 F7       CPL B.7
97    006B: 12 00 72    LCALL SUB_SM_FLOATS
98    006E: 92 E7       MOV ACC.7,C
99    0070: F9          MOV R1,A              ;exponent of result
100   0071: 22          RET
101   ;----------------------------------------------------------------------------
102   0072:            SUB_SM_FLOATS:
103   0072: A2 E7       MOV C,ACC.7
104   0074: 92 D5       MOV PSW.5,C
105   0076: C0 E0       PUSH ACC
106   0078: 65 F0       XRL A,B
107   007A: A2 E7       MOV C,ACC.7
108   007C: 53 F0 7F    ANL B,#01111111B
109   007F: D0 E0       POP ACC
110   0081: 54 7F       ANL A,#01111111B
111   0083: 40 18       JC signs_different
112   0085:            signs_the_same:
113   0085: B5 F0 02    CJNE A,B,different
114   0088: 80 02       SJMP greater_or_equal
115   008A:            different:
116   008A: 40 07       JC less
117   008C:            greater_or_equal:
118   008C: C3          CLR C
119   008D: 95 F0       SUBB A,B
120   008F: A2 D5       MOV C,PSW.5
121   0091: 80 12       SJMP end
122   0093:            less:
123   0093: C3          CLR C
124   0094: C5 F0       XCH A,B
125   0096: 95 F0       SUBB A,B
```

126	0098: A2 D5	MOV C,PSW.5
127	009A: B3	CPL C
128	009B: 80 08	SJMP end
129	009D:	signs_different:
130	009D: 25 F0	ADD A,B
131	009F: A2 E7	MOV C,ACC.7
132	00A1: 92 D2	MOV OV,C
133	00A3: A2 D5	MOV C,PSW.5
134	00A5:	end:
135	00A5: 22	RET
136	;--- end of file ---	

Implementation in code of the <u>multiplication</u>:

- input number: R0 – mantissa of first number, R1 – exponent of first number,
- input number: R2 – mantissa of second number, R3 – exponent of second number,
- output number: R0 – mantissa of result, R1 – exponent of result, OV – result out of the range,
- exemplary value: $[-0.75*2^{-5}]*[+0.625*2^{+1}]$.

```
1      ;****************************************************************
2      ;* Multiplication of floating-point numbers *
3      ;****************************************************************
4      ;mantissa must be normalized, i.e. I after sign bit!
5      ;an exception is mantissa with value of 0.
6
7                    ;c1 EQU 10000101B        ;exponent of first SM number
8                    ;m1 EQU 11100000B        ;mantissa of first SM number
9                    ;c2 EQU 00000001B        ;exponent of second SM number
10                   ;m2 EQU 01010000B        ;mantissa of second SM number
11                   ;first number           m1*2^c1=-0.75*2^-5
12                   ;second number          m2*2^c2=+0.625*2^+1
13                   ;result                 my*2^cy=-0.46875*2^-4=
14                                           ;=-0.9375*2^-5
15                   ;another example of numbers
16     0087          c1 EQU 10000111B        ;exponent of first SM number
17     00D8          m1 EQU 11011000B        ;mantissa of first SM number
18     008C          c2 EQU 10001100B        ;exponent of second SM number
19     00C1          m2 EQU 11000001B        ;mantissa of second SM number
20                   ;first number           m1*2^c1=-0.6875*2^-7
21                   ;second number          m2*2^c2=-0.5078125*2^-12
```

22		;result	my*2^cy=+715/2048*2^−19=
23			;=+1430/2048*2^−20
			=+0.69824*2^−20
24			
25	0000: 78 D8	MOV R0,#m1	
26	0002: 79 87	MOV R1,#c1	
27	0004: 7A C1	MOV R2,#m2	
28	0006: 7B 8C	MOV R3,#c2	
29	0008: 12 00 0D	LCALL MUL_FLOATS	
30			;result in A
31	000B: 80 FE	STOP: SJMP STOP	
32	;---		
33	000D:	MUL_FLOATS:	
34	000D: E8	MOV A,R0	
35	000E: 8A F0	MOV B,R2	
36	0010: 12 00 3A	LCALL MUL_SM_FLOATS	
37	0013: 7C FF	MOV R4,#0FFH	
38	0015:	NORM_MANTISSA:	
39	0015: 20 F6 0A	JB B.6,SKIP	
40	0018: C3	CLR C	
41	0019: 33	RLC A	
42	001A: C5 F0	XCH A,B	
43	001C: 33	RLC A	
44	001D: C5 F0	XCH A,B	
45	001F: 0C	INC R4	
46	0020: 80 F3	SJMP NORM_MANTISSA	
47	0022:	SKIP:	
48	0022: A2 D5	MOV C,PSW.5	
49	0024: 92 F7	MOV B.7,C	
50	0026: A8 F0	MOV R0,B	;mantissa of result
51	0028: E9	MOV A,R1	
52	0029: 8B F0	MOV B,R3	
53	002B: 12 00 4B	LCALL ADD_SM_FLOATS	
54	002E: 20 D2 08	JB OV,SKIP1	
55	0031: 8C F0	MOV B,R4	
56	0033: B2 F7	CPL B.7	
57	0035: 12 00 4B	LCALL ADD_SM_FLOATS	
58	0038: F9	MOV R1,A	;exponent of result
59	0039:	SKIP1:	
60	0039: 22	RET	
61	003A:	MUL_SM_FLOATS:	
62	003A: C0 E0	PUSH ACC	
63	003C: 65 F0	XRL A,B	
64	003E: A2 E7	MOV C,ACC.7	
65	0040: 92 D5	MOV PSW.5,C	

```
 66   0042: 53 F0 7F    ANL B,#01111111B
 67   0045: D0 E0       POP ACC
 68   0047: 54 7F       ANL A,#01111111B
 69   0049: A4          MUL AB
 70   004A: 22          RET
 71   004B:            ADD_SM_FLOATS:
 72   004B: A2 E7       MOV C,ACC.7
 73   004D: 92 D5       MOV PSW.5,C
 74   004F: C0 E0       PUSH ACC
 75   0051: 65 F0       XRL A,B
 76   0053: A2 E7       MOV C,ACC.7
 77   0055: 53 F0 7F    ANL B,#01111111B
 78   0058: D0 E0       POP ACC
 79   005A: 54 7F       ANL A,#01111111B
 80   005C: 50 18       JNC signs_the_same
 81   005E:            signs_different:
 82   005E: B5 F0 02    CJNE A,B,different
 83   0061: 80 02       SJMP greater_or_equal
 84   0063:            different:
 85   0063: 40 07       JC less
 86   0065:            greater_or_equal:
 87   0065: C3          CLR C
 88   0066: 95 F0       SUBB A,B
 89   0068: A2 D5       MOV C,PSW.5
 90   006A: 80 12       SJMP end
 91   006C:            less:
 92   006C: C3          CLR C
 93   006D: C5 F0       XCH A,B
 94   006F: 95 F0       SUBB A,B
 95   0071: A2 D5       MOV C,PSW.5
 96   0073: B3          CPL C
 97   0074: 80 08       SJMP end
 98   0076:            signs_the_same:
 99   0076: 25 F0       ADD A,B
100   0078: A2 E7       MOV C,ACC.7
101   007A: 92 D2       MOV OV,C
102   007C: A2 D5       MOV C,PSW.5
103   007E:            end:
104   007E: 92 E7       MOV ACC.7,C
105   0080: 22          RET
106   ;--- end of file ---
```

Implementation in code of the <u>division</u>:

- input number: R0 – mantissa of first number, R1 – exponent of first number,
- input number: R2 – mantissa of second number, R3 – exponent of second number,
- output number: R0 – mantissa of result, R1 – exponent of result, OV – result out of the range,
- exemplary value: $[-0.5*2^{-7}]/[-0.375*2^{-1}]$.

```
 1    ;********************************************************************
 2    ;* Division of floating-point numbers *
 3    ;********************************************************************
 4    ;mantissa must be normalized, i.e. 1 after sign bit!
 5    ;an exception is mantissa with value of 0.
 6
 7                          ;c1 EQU 00000011B       ;exponent of first SM number
 8                          ;m1 EQU 01010000B       ;mantissa of first SM number
 9                          ;c2 EQU 10000001B       ;exponent of second SM number
10                          ;m2 EQU 10110000B       ;mantissa of second SM number
11                          ;first number          m1*2^c1=+0.625*2^+3
12                          ;second number         m2*2^c2=-0.375*2^-1
13                          ;result                my*2^cy=-106/128*2^+5
14                          ;another example of numbers
15    0087                  c1 EQU 10000111B        ;exponent of first SM number
16    00C0                  m1 EQU 11000000B        ;mantissa of first SM number
17    0084                  c2 EQU 10000100B        ;exponent of second SM number
18    0070                  m2 EQU 01110000B        ;mantissa of second SM number
19                          ;first number          m1*2^c1=-0.5*2^-7
20                          ;second number         m2*2^c2=+0.875*2^-4
21                          ;result                my*2^cy=-73/128*2^-3
22
23    0000: 78 C0           MOV R0,#m1
24    0002: 79 87           MOV R1,#c1
25    0004: 7A 70           MOV R2,#m2
26    0006: 7B 84           MOV R3,#c2
27    0008: 12 00 0D        LCALL DIV_FLOATS
28                                                 ;result in A
29    000B: 80 FE           STOP:SJMP STOP
30    ;-------------------------------------------------------------------
31    000D:                 DIV_FLOATS:
32    000D: E8              MOV A,R0
33    000E: 8A F0           MOV B,R2
34    0010: BA 00 03        CJNE R2,#0,SKIP         ;division by 0!
35    0013: D2 D2           SETB OV
```

```
36   0015: 22            RET
37   0016:               SKIP:
38   0016: 12 00 26      LCALL DIV_SM_FLOATS
39   0019: A2 D5         MOV C,PSW.5
40   001B: 92 E7         MOV ACC.7,C
41   001D: F8            MOV R0,A              ;mantissa of result
42   001E: E9            MOV A,R1
43   001F: 8B F0         MOV B,R3
44   0021: 12 00 39      LCALL SUB_SM_FLOATS
45   0024: F9            MOV R1,A              ;exponent of result
46   0025:               SKIP1:
47   0025: 22            RET
48   0026:               DIV_SM_FLOATS:
49   0026: C0 E0         PUSH ACC
50   0028: 65 F0         XRL A,B
51   002A: A2 E7         MOV C,ACC.7
52   002C: 92 D5         MOV PSW.5,C
53   002E: 53 F0 7F      ANL B,#01111111B
54   0031: D0 E0         POP ACC
55   0033: 54 7F         ANL A,#01111111B
56   0035: 12 00 6F      LCALL FRACTION        ;divide A by B
57   0038: 22            RET
58   0039:               SUB_SM_FLOATS:
59   0039: A2 E7         MOV C,ACC.7
60   003B: 92 D5         MOV PSW.5,C
61   003D: C0 E0         PUSH ACC
62   003F: 65 F0         XRL A,B
63   0041: A2 E7         MOV C,ACC.7
64   0043: 53 F0 7F      ANL B,#01111111B
65   0046: D0 E0         POP ACC
66   0048: 54 7F         ANL A,#01111111B
67   004A: 40 18         JC signs_different
68   004C:               signs_the_same:
69   004C: B5 F0 02      CJNE A,B,different
70   004F: 80 02         SJMP greater_or_equal
71   0051:               different:
72   0051: 40 07         JC less
73   0053:               greater_or_equal:
74   0053: C3            CLR C
75   0054: 95 F0         SUBB A,B
76   0056: A2 D5         MOV C,PSW.5
77   0058: 80 12         SJMP SKIP2
78   005A:               less:
79   005A: C3            CLR C
80   005B: C5 F0         XCH A,B
```

```
81   005D: 95 F0      SUBB A,B
82   005F: A2 D5      MOV C,PSW.5
83   0061: B3         CPL C
84   0062: 80 08      SJMP SKIP2
85   0064:            signs_different:
86   0064: 25 F0      ADD A,B
87   0066: A2 E7      MOV C,ACC.7
88   0068: 92 D2      MOV OV,C
89   006A: A2 D5      MOV C,PSW.5
90   006C:            SKIP2:
91   006C: 92 E7      MOV ACC.7,C
92   006E: 22         RET
93   006F:            FRACTION:
94   006F: 12 00 86   LCALL DENORM
95   0072: 7E 07      MOV R6,#7
96   0074: 7F 00      MOV R7,#0
97   0076:            LOOP:
98   0076: 23         RL A
99   0077: C3         CLR C
100  0078: 95 F0      SUBB A,B
101  007A: 50 02      JNC SKIP3
102  007C: 25 F0      ADD A,B
103  007E:            SKIP3:
104  007E: B3         CPL C
105  007F: CF         XCH A,R7
106  0080: 33         RLC A
107  0081: CF         XCH A,R7
108  0082: DE F2      DJNZ R6,LOOP
109  0084: EF         MOV A,R7
110  0085: 22         RET
111  0086:            DENORM:
112  0086: 7C 00      MOV R4,#00
113  0088:            SHIFT:
114  0088: FD         MOV R5,A
115  0089: C3         CLR C
116  008A: 95 F0      SUBB A,B
117  008C: ED         MOV A,R5
118  008D: 40 05      JC SKIP4
119  008F: C3         CLR C
120  0090: 03         RR A
121  0091: 0C         INC R4
122  0092: 80 F4      SJMP SHIFT
123  0094:            SKIP4:
124  0094: C0 E0      PUSH ACC
125  0096: C0 F0      PUSH B
```

```
126   0098: E9          MOV A,R1
127   0099: 8C F0        MOV B,R4
128   009B: B2 F7        CPL B.7
129   009D: 12 00 39     LCALL SUB_SM_FLOATS
130   00A0: F9           MOV R1,A
131   00A1: D0 F0        POP B
132   00A3: D0 E0        POP ACC
133   00A5: 22           RET
134   ;--- end of file ---
```

For further reading we recommend the books and publications: [Cody 1988, Coonen 1980, Goldberg 1991, IEEE 1985, IEEE 1987, IEEE 2008, Kulisch 2014, Scott 1985 and Sternbenz 1974].

Chapter 6

Limited Quality of Arithmetic Operations

6.1 PRECISION OF NUMBER REPRESENTATION

In the previous chapters, we concentrated on the representation of numbers and on the ways in which the processor performs the four basic arithmetic operations. The aim of this chapter is to make the reader aware of the problem of the finite precision of calculations, which increases with the number of arithmetic operations. It is particularly noticeable in iterative versions of numerical algorithms and in operations on large data structures, e.g. matrices. The causes of the mentioned imperfection are as follows:

- Limited width of a register or memory cell
- Inability to express exactly some numbers on a given basis, e.g. π, 1/3

Let us look at Example 6.1.

Example 6.1: Multiplication and division of decimal fractions 3/4 and 7/2 expressed in BIN format using 4 bits:

$$\frac{3}{4} = 0.110_{BIN} \quad \frac{7}{2} = 11.10_{BIN}$$

a. $\frac{3}{4} \cdot \frac{7}{2} = \frac{21}{8} = 2\frac{5}{8} = 10.101_{BIN} \xrightarrow{\text{but using 4 bits}} \approx 10.10_{BIN}$

b. $\frac{3}{4} : \frac{7}{2} = \frac{3}{4} \cdot \frac{2}{7} = \frac{3}{14} = 0.00110(110)..._{BIN} \xrightarrow{\text{but using 4 bits}} \approx 0.001_{BIN}$

Please note that the result of the product in case (a) of two exactly expressed numbers at the given base (here: $p = 2$) is not exact. The reason is that the result must be written using a limited number of digits (here: 4 bits). The results of the product in case (b) is also approximate, but due to the impossibility of expressing the number 3/14 on the base 2 even if word length is not limited. If even the error of the multiplication can be avoided by using more bits this solution is not satisfying in general case for division

DOI: 10.1201/9781003363286-6

routine. It is obvious that the longer the word length, the higher the precision and the smaller the errors are. However, please remember that numbers are stored in the computer's memory, so if the time of program execution or its size is of primary importance, one should carefully, choose the length of the word to the required precision of numbers, remembering about the accumulation of errors of individual operations. In practice, programmers are developing applications in high-level languages and have several integer and floating-point numeric (see Appendix B for details). The limits of errors are clearly defined and depend on the assumed precision level and the rounding rules. We have a few possibilities, i.e. rounding to the nearest value (the favorite one), toward zero (truncation), toward +inf or toward −inf. In the following discussion, by rounding term, we mean an operation that implements the following rule: if the discarded part is greater than 0.5_{DEC}, increase the preceding digit by one, e.g. 23.17438 -> 23.174, but 23.17458 ->23.175.

 REMEMBER!

If the real number A' is approximated by the number A expressed in the floating-point format of the form $A = M \cdot p^C$, where the mantissa $M = m_0, m_{-1} \cdot m_{-k}$ is composed of k digits in the fractional part, then

- the absolute rounding error $\Delta A = A - A'$ may be positive or negative, and satisfies the inequality $-\frac{1}{2}ulp \leq \Delta A \leq \frac{1}{2}ulp$,
- the absolute truncation error $\Delta A = A - A'$ is always negative, and satisfies the inequality $-ulp < \Delta A \leq 0$, where ulp is the abbreviation for 'units in the last place' and for the assumed number format is equal to $ulp = p^{-k} \cdot p^C$.

Example 6.2: Absolute rounding and truncation errors for the exact number $A' = 12.318_{DEC}$ expressed in floating-point format with k = 2 and k = 3 digits of the fractional part of the mantissa at basis p = 10.

a. **A = round(A′,k)**
   - for $k = 2$ $A = 1.23 \cdot 10^{+1}$
     Absolute rounding error: $\Delta A = A - A' = -0.018$
     We have: $|-0.018| < 0.05$ and $0.5ulp = 0.5 \cdot 10^{-k} \cdot 10^C = 0.5 \cdot 10^{-2} \cdot 10^{+1} = 0.05$
   - for $k = 3$ $A = 1.232 \cdot 10^{+1}$
     Absolute rounding error: $\Delta A = A - A' = +0.002$
     We have $|+0.002| = < 0.005$ and $0.5ulp = 0.5 \cdot 10^{-k} \cdot 10^C = 0.5 \cdot 10^{-3} \cdot 10^{+1} = 0.005$

b. A = **truncate**(A',k)

- for k = 2 A = 1.23 · 10^{+1}
  Absolute truncating error: $\Delta A = A - A' = -0.018$
  We have $-0.1 < -0.018 < 0$, and ulp = $10^{-k} \cdot 10^{C} = 10^{-2} \cdot 10^{+1} = 0.1$
- for k = 3 A = 1.231 · 10^{+1}
  Absolute truncating error: $\Delta A = A - A' = -0.008$

We have $-0.01 < -0.008 < 0$, and ulp = $10^{-k} \cdot 10^{C} = 10^{-3} \cdot 10^{+1} = 0.01$

Exercise 6.1: Determine the rounding and truncation errors of the number A' = 0.0314159_{DEC} expressed in floating-point format with k = 3 and k = 4 digits of the fractional part of the mantissa and p = 10. Check whether the determined errors satisfy the conditions given in the above box.

6.2 ERROR PROPAGATION

In Example 6.1, the case of two decimal input numbers exactly expressed as binary numbers was considered as well. This is not always possible, e.g. try to express a fraction $1/3_{DEC}$ with a base that is not a multiple of 3 is doomed to failure, even assuming an infinite number word length, because $1/3_{DEC} = 0.3333(3)..._{DEC} = 0.0101(01)..._{BIN}$. In general, the input arguments of arithmetic operations may be subject to approximation error. We will check how the introduced inaccuracies propagate by arithmetic operations and have an impact on the error of the result. We will show that it depends on the type of arithmetic operation. One extra assumption was taken. We ignore the error component resulting from the need to approximate the result.

Let A' and B' denote the arguments of arithmetic operations. In a digital machine, they may be written as approximated numbers A and B with absolute error ΔA and ΔB, hence (6.1):

$$A = A' + \Delta A \quad B = B' + \Delta B$$

or

$$\Delta A = A - A' \quad \Delta B = B - B' \tag{6.1}$$

The processor performs an operation on the approximate numbers A and B, so the following formulas hold (6.2):

$$A + B = (A' + \Delta A) + (B' + \Delta B) = A' + B' + (\Delta A + \Delta B) \tag{6.2a}$$

$$A - B = (A' + \Delta A) - (B' + \Delta B) = A' - B' + (\Delta A - \Delta B) \tag{6.2b}$$

$$A \cdot B = (A' + \Delta A) \cdot (B' + \Delta B) = A' \cdot B' + (B' \cdot \Delta A + A' \cdot \Delta B + \Delta A \cdot \Delta B)$$
$$\approx A' \cdot B' + (B' \cdot \Delta A + A' \cdot \Delta B) \tag{6.2c}$$

Let us derive the formula for division. If we assume, as an alternative (6.3):

$$\frac{A}{B} = \frac{A' + \Delta A}{B' + \Delta B} = \frac{A'}{B'} + R \tag{6.3}$$

hence (6.4):

$$R = \frac{A' + \Delta A}{B' + \Delta B} - \frac{A'}{B'} = \frac{(A' + \Delta A) \cdot B' - A' \cdot (B' + \Delta B)}{(B' + \Delta B) \cdot B'}$$
$$= \frac{A' \cdot B' + \Delta A \cdot B' - A' \cdot B' - A' \cdot \Delta B}{(B' + \Delta B) \cdot B'} = \frac{\Delta A \cdot B' - A' \cdot \Delta B}{(B' + \Delta B) \cdot B'} \tag{6.4}$$
$$\approx \frac{\Delta A \cdot B' - A' \cdot \Delta B}{B' \cdot B'} = \frac{\Delta A}{B'} - \frac{A' \cdot \Delta B}{B' \cdot B'}$$

and finally (6.5):

$$\frac{A}{B} = \frac{A' + \Delta A}{B' + \Delta B} = \frac{A'}{B'} + \frac{\Delta A \cdot B' - A' \cdot \Delta B}{(B' + \Delta B) \cdot B'} \approx \frac{A'}{B'} + \left(\frac{\Delta A}{B'} - \frac{A' \cdot \Delta B}{B' \cdot B'} \right) \tag{6.5}$$

In summary, the arithmetic operations listed are subject to absolute errors (6.6):

$$\Delta_{A+B} = \Delta A + \Delta B \quad \Delta_{A-B} = \Delta A - \Delta B \tag{6.6a}$$

$$\Delta_{A \cdot B} \approx A' \cdot \Delta B + B' \cdot \Delta A \quad \Delta_{A/B} \approx \frac{\Delta A}{B'} - \frac{A' \cdot \Delta B}{B' \cdot B'} \tag{6.6b}$$

The simplified formulas for multiplication and division error is valid when $A \gg \Delta A$ and $B \gg \Delta B$, otherwise the exact full formulas should be used.

Similarly to the absolute error, the formula for relative error can be derived, defined in general (6.7):

$$\delta X = \frac{X - X'}{X'} = \frac{\Delta X}{X'} \tag{6.7}$$

Accordingly, we get (6.8):

$$\delta_{A+B} = \frac{\Delta A + \Delta B}{A' + B'} = \frac{A'}{A' + B'} \cdot \frac{\Delta A}{A'} + \frac{B'}{A' + B'} \cdot \frac{\Delta B}{B'}$$

$$= \frac{A'}{A' + B'} \cdot \delta A + \frac{B'}{A' + B'} \cdot \delta B \qquad (6.8a)$$

$$\delta_{A-B} = \frac{\Delta A - \Delta B}{A' - B'} = \frac{A'}{A' - B'} \cdot \frac{\Delta A}{A'} - \frac{B'}{A' - B'} \cdot \frac{\Delta B}{B'}$$

$$= \frac{A'}{A' - B'} \cdot \delta A - \frac{B'}{A' - B'} \cdot \delta B \qquad (6.8b)$$

$$\delta_{A \cdot B} \approx \frac{A' \cdot \Delta B + B' \cdot \Delta A}{A' \cdot B'} = \frac{\Delta A}{A'} + \frac{\Delta B}{B'} = \delta A + \delta B \qquad (6.8c)$$

$$\delta_{A/B} \approx \left(\frac{\Delta A}{B'} - \frac{A' \cdot \Delta B}{B' \cdot B'} \right) \cdot \frac{B'}{A'} = \frac{\Delta A}{A'} - \frac{\Delta B}{B'} = \delta A - \delta B \qquad (6.8d)$$

We leave it to the reader to derive the exact relations for the relative error of multiplication and division. After analyzing the formulas obtained, the following observations arise:

- the absolute error of addition and subtraction depends only on the error of the approximations of the arguments,
- the relative error of the result of multiplication and division depends only on the error of the approximations of the arguments,
- the relative error of the result of subtraction is greater the smaller the difference of the arguments.

Potentially worrying is the last observation, which shows that the relative error of subtraction can many times exceed the relative errors of the arguments! Such a situation is shown in Example 6.3. Please compare the result errors in cases (a) and (b).

Example 6.3: Relative errors of addition and subtraction of two decimal numbers A' and B', where

a. $A' = 100$, $\Delta A = 1$, $B' = 99$, $\Delta B = 0.9$

hence $\delta A = 0.01 = 1\%$, $\delta B \approx 0.01 = 1\%$

$$\delta_{A+B} = \frac{\Delta A + \Delta B}{A' + B'} = \frac{1 + 0.9}{100 + 99} \approx \frac{1}{100} = 1\% < 1\% + 1\%$$

$$\delta_{A-B} = \frac{\Delta A - \Delta B}{A' - B'} = \frac{1 - 0.9}{100 - 99} = \frac{1}{10} = 10\%$$

b. $A' = 100$, $\Delta A = 1$, $B' = 99.9$, $\Delta B = 0.9$

hence $\delta A = 0.01 = 1\%$, $\delta B \approx 0.01 = 1\%$

$$\delta_{A+B} = \frac{\Delta A + \Delta B}{A' + B'} = \frac{1 + 0.9}{100 + 99.9} = \frac{1.9}{199.9} \approx \frac{1}{100} = 1\% < 1\% + 1\%$$

$$\delta_{A-B} = \frac{\Delta A - \Delta B}{A' - B'} = \frac{1 - 0.9}{100 - 99.9} = \frac{0.1}{0.1} = 1 = 100\%!$$

Exercise 6.2: Determine the relative error of addition, subtraction, multiplication and division of two decimal numbers A' and B', where A'= 543, $\Delta A = 2$, B'= 398 and $\Delta B = 3$.

Example 6.4: Numbers $A' = \frac{5}{6}$, $B' = \frac{2}{7}$ as decimal fractions to three decimal places and the errors of arithmetical operations on approximate numbers to which rounding has been applied:

    a. A + B
    b. A − B
    c. A · B
    d. A/B

$A = 0.833_{DEC}$ $B = 0.285_{DEC}$

$$\Delta A = A - A' = \frac{833}{1000} - \frac{5}{6} = \frac{2499}{3000} - \frac{2500}{3000} = -\frac{1}{3000}$$

$$\Delta B = B - B' = \frac{285}{1000} - \frac{2}{7} = \frac{1995}{7000} - \frac{2000}{7000} = -\frac{5}{7000}$$

a)

$$\Delta A + \Delta B = -\frac{1}{3000} + \left(-\frac{5}{7000}\right) = -\frac{7 + 15}{21000}$$

$$= -\frac{22}{21000} = -0.0010(476190)... \approx -0.001$$

Checking:

$A + B = 0.833 + 0.285 = 1.118$

$$A' + B' = \frac{5}{6} + \frac{2}{7} = \frac{35}{42} + \frac{12}{42} = \frac{47}{42} = 1\frac{5}{42}$$

$$\Delta_{A+B} = (A + B) - (A' + B') = 1\frac{118}{1000} - 1\frac{5}{42} = \frac{4956}{42000} - \frac{5000}{42000} = -\frac{22}{21000}$$
$$= -0.0010(476190)... \approx -0.001$$

b)

$$\Delta A - \Delta B = -\frac{1}{3000} - \left(-\frac{5}{7000}\right) = \frac{15 - 7}{21000}$$

$$= -\frac{8}{21000} = -0.000(380952)... \approx -0.0004$$

Checking:

$A-B = 0.833 - 0.285 = 0.548$

$$A' - B' = \frac{5}{6} - \frac{2}{7} = \frac{35}{42} - \frac{12}{42} = \frac{23}{42}$$

$$\Delta_{A-B} = (A - B) - (A' - B') = \frac{548}{1000} - \frac{23}{42} = \frac{23016}{42000} - \frac{23000}{42000}$$
$$= -\frac{8}{21000} - 0.000(380952)... \approx -0.0004$$

c)

$$B'\cdot\Delta A + A'\cdot\Delta B = \frac{2}{7}\cdot\left(-\frac{1}{3000}\right) + \frac{5}{6}\cdot\left(-\frac{5}{7000}\right) = -\frac{1}{1000}\cdot\left(\frac{2}{21} + \frac{25}{42}\right)$$

$$= -\frac{1}{1000}\cdot\frac{25+4}{42} = -0.0006947...$$

Checking:

$A \cdot B = 0.833 \cdot 0.285 = 0.237405$

$$A' \cdot B' = \frac{5}{6} \cdot \frac{2}{7} = \frac{10}{42}$$

$$\Delta_{A \cdot B} = (A \cdot B) - (A' \cdot B') = \frac{237405}{1000000} - \frac{10}{42} = \frac{997101}{4200000} - \frac{1000000}{4200000}$$

$$= -\frac{2899}{4200000} = -0.000690(238095)...$$

d)

$$\frac{\Delta A}{B'} - \frac{A' \cdot \Delta B}{B' \cdot B'} = -\frac{1}{3000} \cdot \frac{7}{2} - \frac{5}{6} \cdot \left(-\frac{5}{7000}\right) \cdot \frac{7}{2} \cdot \frac{7}{2} = -\frac{7}{6000} + \frac{175}{24000} = \frac{175 - 28}{24000}$$

$$= \frac{147}{24000} = 0.006125$$

Checking:

$$A/B = 0.833/0.285 = \frac{833}{285} = 2\frac{263}{285}$$

$$A'/B' = \frac{5}{6} \cdot \frac{7}{2} = \frac{35}{12} = 2\frac{11}{12}$$

$$\Delta_{A/B} = (A/B) - (A'/B') = 2\frac{263}{285} - 2\frac{11}{12} = \frac{3156}{3420} - \frac{3135}{3420} = \frac{21}{3420}$$
$$= 0.00(614035087719298245)... \approx 0.006$$

Exercise 6.3: Given decimal fractions A' = 3/16, B' = 5/8. Write them down in the form of binary fractions A and B to three decimal places, and then determine the errors of the following arithmetic operations:

  a. A + B
  b. A − B
  c. A · B
  d. A/B

Show the correctness of the formulas for the absolute errors of the operations listed above.

The error propagation and amplification can be a cause of serious final errors, particularly if mathematical operations are done many times. A lot of scientific algorithms are based on iterative solution of task that requires repetition of many laps in the loop. Another example of numerous operations is matrix algebra used to describe many research problems requiring solving systems of equations (we need matrix inversion and multiplication as well). The best practice is to limit the necessary amount of calculations and assume format of numbers with some extra significant digits. Fortunately, in the most cases, the double format proposed by IEEE 754 standard seems to be accurate.

The deeper understanding how precise the result of a floating-point calculation is and which operations introduce the most significant errors is not easy work and generally is out of a scope and main objectives of this simple book. For further reading, some excellent papers can be recommended. In [Martel 2006], general method of assessment based on abstract interpretation was discussed from theoretical point of view. Nevertheless, it can be too hard for students without understanding of sophisticated math. Another interesting approach is the CADNA library [Jézéquel 2008], which allows you to estimate the propagation of rounding errors using a probabilistic approach. With CADNA, you can control the numerical quality of any simulation program. In addition, numerical debugging of user code can be performed by detecting any instabilities that may occur during working.

Remarks

1 IT APPLIES TO OPERATIONS ON NUMBERS IN FIXED-POINT FORMAT

Analyzing the length of the resulting code, it seems that in many cases a shorter code can be obtained by reducing it to the BIN form. Then it is enough to use the appropriate arithmetic instructions ADD, SUBB, MUL and DIV. In the final phase the result should be converted into the desired output form. For which formats and actions is such a procedure worthwhile? Unfortunately, or rather lackily, the Reader must find the answer to this question himself.

2 IT APPLIES TO OPERATIONS ON NUMBERS IN FLOATING-POINT FORMAT

The format of floating-point numbers adopted in Chapter 5 is characterized by low precision, so you should be aware that the usefulness of the subroutines contained therein is limited. They were aimed to provide, straightforward as possible, the link between theoretical considerations given in Chapter 4 and programming practice. Maybe it will facilitate the understanding of coding and encourage the Reader to search for own programmatic implementations. I know that is a challenge but giving a lot of satisfaction. Adopting, for example, the single-precision format, as recommended by the IEEE standard, would certainly result in a more extensive, and at the same time less readable, code of subroutines, which could effectively discourage an attempt to analyze them.

3 GENERAL REMARK

Due to primary educational purpose of this book many advanced topics were omitted, unfortunately. We could spend many times discussing, for

DOI: 10.1201/9781003363286-7

example the issues of algorithms implementation in assembly code of various microprocessors starting with simple 8bits architecture and finish with 64bits one. We did not discuss the benefits of classical math library delivering the ready-to-use arithmetic functions working with fixed and floating formats for many CPUs. The reader could be also curious what are the CORDIC method improvements and limitations and many more? People interesting in hardware or mixed code/hardware realization of arithmetic, what is common solutions met in the Graphical Processing Units (GPU chips used in graphical cards or laptop chipsets) may feel a little bit disappointed. The author is fully aware of these limitations. If you have reached this point of book, you are probably also unsatisfied and looking for further knowledge. I congratulate you on your persistence, but you must go on your own way, explore and collect experience by reading excellent books, e.g. [Brent 2010], [Koren 2002] and papers [Volder 1959], [Li 2016] and practice and practice more ...

Best regards
Author

References

BOOK AND JOURNALS

Augarten S., *Bit by bit: An illustrated history of computers*, Unwin Paperbacks, London 1985.

Baer J.L., *Microprocessor architecture: From simple pipelines to chip multiprocessors*, Cambridge University Press, New York, 2010.

Biernat J., *Architektura układów arytmetyki resztowej (en. Architecture of residual arithmetic systems)*, Akademicka Oficyna Wydawnicza EXIT, Warszawa, 2007, in Polish.

Bindal A., *Fundamentals of computer architecture and design*, 2nd edition, Springer, Cham, 2019.

Blaauw G.A., Brooks F.P.J.R., *Computer architecture: Concepts and evolution*, Addison-Wesley, Reading, 1997.

Boot A.D., A signed binary multiplication technique, *Journal of Applied Mathematics*, 4(2/1951) 1951, pp. 236–240.

Brent R., Zimmermann P., *Modern computer arithmetic*, Cambridge University Press, Cambridge, 2010.

Ceruzzi P., *A history of modern computing*, MIT Press, Cambridge, 1998.

Cherkauer B., Friedman E., A hybrid radix-4/radix-8 low power, high speed multiplier architecture for wide bit widths, *IEEE International Symposium on Circuits and Systems*, 1996, pp. 53–56.

Cody W.J., Floating-point standards – theory and practice, *Reliability in computing: the role of interval methods in scientific computing*, pp. 99–107, Academic Press, Boston, 1988.

Coonen J., An implementation guide to a proposed standard for floating-point arithmetic, *Computer*, 13, pp. 68–79.

Efstathiou C., Vergos H., Modified booth 1's complement and modulo 2n-1 multipliers, *The 7th IEEE International Conference on Electronics Circuits and Systems*, 2000, pp. 637–640.

Flores I., *The logic of computer arithmetic*, Englewood Cliffs, Prentice-Hall Inc., New York, 1962.

Goldberg D., *What every computer scientist should know about floating-point arithmetic*, Computing Surveys, Association for Computing Machinery Inc., March 1991.

Gryś S., Signed multiplication technique by means of unsigned multiply instruction, *Computers and Electrical Engineering*, 37, 2011, pp. 1212–1221, doi:10.101 6/j.compeleceng.2011.04.004.

Gryś S., Minkina W., O znaczeniu odwrotnej notacji polskiej dla rozwoju technik informatycznych (en. On the importance of reverse polish notation for the development of computer science), *Pomiary Automatyka Robotyka, ISSN 1427-9126, R. 24*, No. 24(2) 2020, pp. 11–16, doi: 10.14313/PAR_236/11, in Polish.

Hamacher C., Vranesic Z. et al., *Computer organization and embedded systems, 6th edition*, McGraw-Hill, New York, 2012.

Hohl W., Hinds Ch., *ARM assembly language: Fundamentals and techniques – 2nd edition*, CRC Press, Boca Raton, 2015.

Hwang K., *Computer arithmetic: Principles, architecture and design*, John Wiley & Sons Inc., New York, 1979.

IEEE Standard 754-1985 for binary floating-point arithmetic, ANSI/IEEE 1985.

IEEE Standard 854-1987 for radix-independent floating-point arithmetic, ANSI/IEEE 1987.

IEEE Standard 754-2008 for floating-point arithmetic, ANSI/IEEE 2008.

Irvine K., *Assembly language for x86 processors 7th edition*, Pearson, Upper Saddle River, 2014.

Jézéquel F., Chesneaux J-M., CADNA: A library for estimating round-off error propagation, *Computer Physics Communications*, 178, 2008, pp. 933–955, 10.1016/j.cpc.2008.02.003.

Koren I., *Computer arithmetic algorithms*, Prentice-Hall, Englewood Cliffs, New Jersey, 1993.

Koren I., *Computer arithmetic algorithms, 2nd edition*, A.K. Peters, Natick Massachusetts, 2002.

Kulisch U., *Advanced arithmetic for the digital computer: Design of arithmetic units*, Springer Science & Business Media, Wien, 2012.

Kulisch U., Miranker W., *Computer arithmetic in theory and practice*, Academic Press, New York, 2014.

Li J., Fang J., Li J., Zhao Y., Study of CORDIC algorithm based on FPGA, *2016 Chinese Control and Decision Conference (CCDC)*, pp. 4338–4343, doi: 10.1109/CCDC.2016.7531747.

Mano M., *Computer system architecture*, Prentice Hall, Englewood Cliffs, 1993.

Mano M., *Computer system architecture, 3rd edition*, Pearson Education, London, 2008.

Martel M., Semantics of roundoff error propagation in finite precision calculations, *Higher-Order Symbolic Computation*, 19(7–30) 2006, doi: 10.1007/s10990-006-8608-2.

Matula D., Kornerup P., Finite precision rational arithmetic: Slash number systems, *IEEE Transactions on Computers*, C-34(1) 1985, pp. 3–18.

McCartney S., *ENIAC: The triumphs and tragedies of the world's first computer*, Walker and Company, New York, 1999.

McSorley O.L., High speed arithmetic in binary computers, *Proceedings of IRE*, January 1961, pp. 67–91.

Metzger P., *Anatomia PC (en. Anatomy of PC)*, wyd. XI, Helion, Gliwice 2007, in Polish.

Mollenhoff C., *Atanasoff: The forgotten father of the computer*, Iowa State University Press, Ames, 1988.

Null L., Lobur J., *The essentials of computer organization and architecture*, John and Barlett Publishers, Burlington, 2018.

Omondi A.R., *Computer arithmetic systems, algorithms, architecture and implementations*, Series in Computer Science Prentice-Hall International, Englewood Cliffs, New York, 1994.

Pankiewicz S., *Arytmetyka liczb zapisywanych w systemach niedziesiętnych (en. Arithmetic of numbers in non-decimal systems)*, Politechnika Śląska, Gliwice, 1985, in Polish.

Parhami B., *Computer arithmetic: Algorithms and hardware designs*, Oxford University Press, New York, 2010.

Patterson D., Hennessy J., *Computer organisation and design: The hardware/software interface, 5th edition*, Morgan Kaufmann, Oxford, 2014.

Pochopień B., *Arytmetyka komputerowa (en. Computer arithmetic)*, Akademicka Oficyna Wydawnicza EXIT, Warsaw, 2012, in Polish.

Pollachek H., *Before the ENIAC, IEEE annals of the history computing 19*, June 1997, pp. 25–30.

Richards R., *Arithmetic operations in digital computers*, Princeton, D.Van Nostrand, New York, 1955.

Ruszkowski P., Witkowski J., *Architektura logiczna i oprogramowanie prostych mikroukładów kalkulatorowych (en. Logical architecture and software for simple calculator microcircuits)*, PWN, Warsaw, 1983, in Polish.

Schmid H., *Decimal arithmetic*, John Wiley & Sons Inc., New York, 1979.

Scott N., *Computer number systems and arithmetic*, Prentice-Hall, Englewood Cliffs, New York, 1985.

Seidel P., McFearin L., Matula D., *Binary multiplication radix-32 and radix-256*, 15th Symposium on Computer Arithmetic, 2001, pp. 23–32.

Stallings W. *Computer organization and architecture, designing for performance*, 8th edition, Pearson Education, Upper Sadle River, 2008.

Sternbenz P.H., *Floating-point computation*, Prentice-Hall, Englewood Cliffs, New York, 1974.

Swartzlander E., Alexopoulos A., The sign/logarithm number systems, *IEEE Transactions on Computers*, C-24(12) 1975, pp. 1238–1242.

Swartzlander E. (ed.), *Computer arithmetic*, vol. I, World Scientific, 2015, ISBN 978-981-4651-56-1, 10.1142/9476.

Tietze U., Schenk Ch., Gamm E., *Electronic circuits: Handbook for design and applications*, 2nd edition, Springer, Berlin, 2002.

Vitali A., *Coordinate rotation digital computer algorithm (CORDIC) to compute trigonometric and hyperbolic functions*, DT0085 Design tip, ST Microelectronics, 2017.

Vladutiu M., *Computer arithmetic: Algorithms and hardware implementations*, Springer-Verlag, Berlin Heidelberg 2012.

Volder J., The CORDIC computer technique, IRE-AIEE-ACM '59 (Western), pp. 257–261, 10.1145/1457838.1457886.

Wiki 2022, CORDIC, from Wikipedia, https://en.wikipedia.org/wiki/CORDIC.

Appendices

Numerical range for BIN, 2's and SM formats for assumed n and m, where n is the number of bits of the integer part (quotient), m is the number of bits of the fractional part.

Table A.1 Range for n + m = 8 Bits

n	m	Resolution 2^{-m}	Range for decimal format		
			BIN	2's	SM
8	0	1.0	0:255	−128:127	−127:127
7	1	0.5	0:127.5	−64:63.5	−63.5:63.5
6	2	0.25	0:63.75	−32:31.75	−31.75:31.75
5	3	0.125	0:31.875	−16:15.875	−15.875:15.875
4	4	0.0625	0:15.9375	−8:7.9375	−7.9375:7.9375
3	5	0.03125	0:7.96875	−4:3.96875	−3.96875:3.96875
2	6	0.015625	0:3.984375	−2:1.984375	−1.984375: 1.984375
1	7	0.0078125	0:1.9921875	−1:0.9921875	−0.9921875: 0.9921875
0	8	0.00390625	0:0.99609375	−0.5:0.49609375	−0.49609375:0.49609375

Table A.2 Range for n + m = 16 Bits

n	m	Resolution 2^{-m}	Range for decimal format		
			BIN	2's	SM
16	0	1.0	0:65535	−32768:32767	−32767:32767
15	1	0.5	0:32767.5	−16384:16383.5	−16383.5:16383.5
14	2	0.25	0:16383.75	−8192:8191.75	−8191.75:8191.75
13	3	0.125	0:8191.875	−4096:4095.875	−4095.875:4095.875
12	4	0.0625	0:4095.9375	−2048:2047.9375	−2047.9375:2047.9375
11	5	0.03125	0:2047.96875	−1024:1023.96875	−1023.96875:1023.96875
10	6	0.015625	0:1023.984375	−512:511.984375	−511.984375:511.984375
9	7	0.0078125	0:511.9921875	−256:255.9921875	−255.9921875:255.9921875
8	8	0.00390625	0:255.99609375	−128:127.99609375	−127.99609375:127.99609375
7	9	0.001953125	0:127.9988046875	−64:63.998046875	−63.998046875:63.998046875
6	10	0.0009765625	0:63.9990234375	−32:31.9990234375	−31.9990234375:31.9990234375
5	11	0.00048828125	0:31.99951171875	−16:15.99951171875	−15.99951171875:15.99951171875
4	12	0.000244140625	0:15.999755859375	−8:7.999755859375	−7.999755859375:7.999755859375
3	13	0.0001220703125	0:7.9998779296875	−4:3.9998779296875	−3.9998779296875:3.9998779296875
2	14	0.00006103515625	0:3.99993896484375	−2:1.99993896484375	−1.99993896484375:1.99993896484375
1	15	0.000030517578125	0:1.999969482421875	−1:0.999969482421875	−0.999969482421875:0.999969482421875
0	16	0.0000152587890625	0:0.9999847412109375	−0.5:0.4999847412109375	−0.4999847412109375:0.4999847412109375

APPENDIX B. NUMERICAL DATA TYPES IN SOME HIGH-LEVEL LANGUAGES

Table B.1 Integer Types of Numeric Variables

Signed numbers		Delphi Pascal	C/C++	Java	Microsoft Visual Basic
Range	Bytes	Name			
−128... +127	1	shortInt	signed char	byte	SByte
−32768... +32767	2	SmallInt	short short int signed short signed short int	short	short Int16
−2147483648... +2147483647	4	LongInt	long long int signed long signed long int	int	integer Int32
−9 223 372 036 854 775 808 ... + 9 223 372 036 854 775 807	8	Int64	long long long long int signed long long signed long long int	long	long Int64
Unsigned numbers		**Delphi Pascal**	**C/C++**	**Java**	**Microsoft Visual Basic**
Range	*Bytes*	*Name*			
0... 255	1	Byte	unsigned char	–	byte
0... 65535	2	Word	unsigned short unsigned short int	char	UShort UInt16
0... 4 294 967 295	4	Longword Cardinal	unsigned long unsigned long int	Int (for SE8 and higher releases	UInteger UInt32
0 ... 18 446 744 073 709 551 615	8	–	unsigned long long unsigned long long int	Long (for SE8 and higher releases)	Ulong UInt64

Table B.2 Real Numbers

Range (normalized numbers)	Bytes	Precision (significant decimal digits)	Name				
			IEEE-P754	Delphi Pascal	C/C++	Java	Microsoft Visual Basic
±(1.18e−38...3.40e38)	4	7–8	single	Single	float	Float	Single
±(2.23e−308...1.79e308)	8	15–16	double	Double	double	double	Double
±(3.37e−4932...1.18e4932)	10	19	double extended	Extended	long double	–	–
±(6.1e−5 ... 65504)	2	3–4	half (binary16)	–	half	–	–
±(3.37e−4932...1.18e4932)	16	34	quadruple (binary128)	–	_Float128	–	–

APPENDIX C. SOLUTIONS TO EXERCISES

Exercise 1.1: Tip. From dependencies $a \cup a = a$, $a \cap a = a$ and $a \oplus a = 0$ we have:

$$
\begin{array}{c}
\quad a_3 \quad a_2 \quad 0 \quad 1 \\
\cup \ a_3 \quad 1 \quad a_1 \quad 1 \\
\hline
\quad a_3 \quad 1 \quad a_1 \quad 1
\end{array}
\qquad
\begin{array}{c}
\quad a_3 \quad a_2 \quad 0 \quad 0 \\
\cap \ 1 \quad a_2 \quad 0 \quad a_0 \\
\hline
\quad a_3 \quad a_2 \quad 0 \quad 0
\end{array}
\qquad
\begin{array}{c}
\quad 0 \quad 1 \quad a_1 \quad a_0 \\
\oplus \ a_3 \quad 1 \quad 1 \quad a_0 \\
\hline
\quad a_3 \quad 0 \quad /a_1 \quad 0
\end{array}
$$

Exercise 2.1: $p = 2$, $1.11_{BIN} = 1.75_{DEC} = 2^1 - 2^{-2}$,

$p = 10$, $9.99_{DEC} = 10^1 - 10^{-2}$,

$p = 16$, $F.FF_{HEX} = 15 = 255/256 = 16^1 - 16^{-2}$.

Exercise 2.2:

a. $246.5_{DEC} \rightarrow 11110110.1000_{BIN} \rightarrow F6.8_{HEX}$
b. $3E.4_{HEX} \rightarrow 00111110.0100_{BIN} \rightarrow 62.25_{DEC}$
c. $10110011.0010_{BIN} \rightarrow B3.2_{HEX} \rightarrow 179.125_{DEC}$

Exercise 2.3:

a. $0.63_{DEC} \rightarrow \approx 0.10100000_{BIN} \rightarrow 0.A0_{HEX}$
b. $11/9_{DEC} = (1 + 2/9)_{DEC} \rightarrow \approx 1.00111000_{BIN} \rightarrow 1.38_{HEX}$
c. $3/5_{DEC} \rightarrow \approx 0.10011001_{BIN} \rightarrow \approx 0.99_{HEX}$
d. $1/128_{DEC} \rightarrow 0.00000010_{BIN} \rightarrow 0.02_{HEX}$

Exercise 2.4:

a. $479.12_{DEC} \rightarrow 010001111001.00010010_{P\text{-}BCD} \rightarrow 00000100000001110000$
   $1001.000000010000010_{UP\text{-}BCD}$
b. $0.03_{DEC} \rightarrow 0000.00000011_{P\text{-}BCD} \rightarrow 00000000.000000000000011_{UP\text{-}BCD}$
c. $8.9_{DEC} \rightarrow 1000.1001_{P\text{-}BCD} \rightarrow 00001000.00001001_{UP\text{-}BCD}$
d. $123_{DEC} \rightarrow 000100100011_{P\text{-}BCD} \rightarrow 00000001000000100000011_{UP\text{-}BCD}$

Exercise 2.5:

a. $361.82_{DEC} \rightarrow 3336312E3832_{(HEX) \text{ as ASCII}}$
b. $36.18_{DEC} \rightarrow 33362E3138_{(HEX) \text{ as ASCII}}$
c. $0.45_{DEC} \rightarrow 302E3435_{(HEX) \text{ as ASCII}}$
d. $97.1_{DEC} \rightarrow 39372E31_{(HEX) \text{ as ASCII}}$

Exercise 2.6:

a. $+23.5_{DEC} \rightarrow \underline{0}10111.1_{SM}$
b. $+17.3_{DEC} \rightarrow \underline{0}10001.0100\ldots_{SM}$
c. $-11.25_{DEC} \rightarrow \underline{1}1011.01_{SM}$
d. $-1_{DEC} \rightarrow \underline{1}1_{SM}.$

Exercise 2.7:

a. $+3.125_{DEC} \rightarrow 011.001_{2's}$
b. $-17.5_{DEC} \rightarrow 101110.1_{2's}$
c. $-1_{DEC} \rightarrow 1_{2's}$
d. $+1_{DEC} \rightarrow 01_{2's}$

Exercise 3.1:

a.

$$
\begin{array}{r}
11.011101_{BIN} \\
+\ 00.101101_{BIN} \\
\hline
1\ 00.001010_{BIN}
\end{array}
$$

b.

$$
\begin{array}{r}
1101.0111_{BIN} \\
+\ 1010.1100_{BIN} \\
\hline
1\ 1000.0011_{BIN}
\end{array}
$$

Exercise 3.2:

a.

$$
\begin{array}{r}
11.011100_{BIN} \\
-\ 01.101011_{BIN} \\
\hline
01.110001_{BIN}
\end{array}
$$

b.

$$
\begin{array}{r}
1\ 0101.1011_{BIN} \\
-\ 1010.1101_{BIN} \\
\hline
1010.1110_{BIN}
\end{array}
$$

Exercise 3.3:

a. $A = 42.5_{DEC}$ $B = 68_{DEC}$ $p = 2$

$A = 101010.1_{BIN}$ $B = 1000100.0_{BIN}$ $\bar{B} = 0111011.1$ $\bar{\bar{B}} = 0111100.0$

$$
\begin{array}{ll}
\quad 0101010.1 & A \\
+\ 0111100.0 & \bar{\bar{B}} \\
\hline
\cancel{0}\ 1100110.1 & A + \bar{\bar{B}} \\
\qquad \Downarrow & \\
\underline{1}\ 0011001.1 & -\left(\overline{\overline{A + \bar{\bar{B}}}}\right)
\end{array}
$$

$$
\begin{array}{ll}
\quad 0101010.1 & A \\
+\ 0111011.1 & \bar{B} \\
\hline
\cancel{0}\ 1100110.0 & A + \bar{B} \\
\qquad \Downarrow & \\
\underline{1}\ 0011001.1 & -(\overline{A + \bar{B}})
\end{array}
$$

b. $A = 75_{DEC}$ $B = 13_{DEC}$ $p = 10$ $\bar{B} = 86$ $\bar{\bar{B}} = 87$

```
      75  A                    75  A
   + 87  B̿                  + 86  B̄
  ─────────                 ─────────
  1  62  A + B̿             1  61  A+B̄
      ⇓                    + 01  p⁻ᵐ
                           ─────────
  0  62  +(A + B̿)          0  62  +(A + B̄)
```

$$\begin{array}{r} 75 \ \ A \\ +\ 87 \ \ \bar{\bar{B}} \\ \hline 1 \ \ 62 \ \ A + \bar{\bar{B}} \\ \Downarrow \\ \underline{0} \ \ 62 \ \ +(A + \bar{\bar{B}}) \end{array} \qquad \begin{array}{r} 75 \ \ A \\ +\ 86 \ \ \bar{B} \\ \hline 1 \ \ 61 \ \ A+\bar{B} \\ +\ 01 \ \ p^{-m} \\ \hline \underline{0} \ \ 62 \ \ +(A + \bar{B}) \end{array}$$

Exercise 3.4:

a.

$$\begin{array}{r} 11.11_{BIN} \\ *\ 0.101_{BIN} \\ \hline 1111 \\ 0000 \\ 1111 \\ *\ \ 0000 \\ \hline 10.01011_{BIN} \end{array}$$

b.

$$\begin{array}{r} 1.001_{NKD} \\ *\ 0110_{NKD} \\ \hline 0000 \\ 1001 \\ 1001 \\ *\ \ 0000 \\ \hline 0110.110_{NKD} \end{array}$$

Exercise 3.5:

a)

$$\begin{array}{l} 110010_{BIN} \\ -\ \ 001101_{BIN} \\ \hline \ \ 100101 \quad >0 \Rightarrow \text{quotient} = 1_{DEC} \\ -\ \ 001101 \\ \hline \ \ 011000 \quad >0 \Rightarrow \text{quotient} = 2_{DEC} \\ -\ \ 001101 \\ \hline \ \ 001011 \quad >0 \Rightarrow \text{quotient} = 3_{DEC} = 011_{BIN} \\ -\ \ 010010 \\ \hline \ \ 111001 \quad <0 \Rightarrow \text{reminder} = 101_{BIN} \end{array}$$

b)

$$\begin{array}{l} \ \ \approx 11.1..._{BIN} \\ \hline 110010_{BIN} \quad :1101_{BIN} \\ -01101 \\ \hline \ \ 11000 \\ -\ \ 01101 \\ \hline \ \ 10110 \\ -\ \ 01101 \\ \hline \ \ \ 1001 \\ \quad \quad ... \end{array}$$

or

$$\begin{array}{l} \quad \quad \quad 11_{BIN} \quad - \text{quotient} \\ \hline 110010_{BIN} \quad :1101_{BIN} \\ -01101 \\ \hline \ \ 11000 \\ -\ \ 01101 \\ \hline \ \ 01011 \quad - \text{reminder} \end{array}$$

Exercise 3.6:

$$0001 \xleftarrow{1} 0011$$
$$- \quad 1001$$
$$\overline{1000} < 0 \Rightarrow \text{quotient} = 0$$
$$+ \quad 1001$$
$$\overline{0001}$$
$$\downarrow$$
$$0010 \xleftarrow{0} 011$$
$$- \quad 1001$$
$$\overline{1001} < 0 \Rightarrow \text{iloraz} = 00$$
$$+ \quad 1001$$
$$\overline{0010}$$
$$\downarrow$$
$$0100 \xleftarrow{0} 11$$
$$- \quad 1001$$
$$\overline{1011} < 0 \Rightarrow \text{quotient} = 000$$
$$+ \quad 1001$$
$$\overline{0100}$$
$$\downarrow$$
$$1001 \xleftarrow{1} 1$$
$$- \quad 1001$$
$$\overline{0000} \geq 0 \Rightarrow \text{quotient} = 0001$$
$$\downarrow$$
$$0001 \xleftarrow{1}$$
$$- \quad 1001$$
$$\overline{1000} < 0 \Rightarrow \text{quotient} = 00010$$
$$+ \quad 1001$$
$$\overline{\text{reminder} = \quad 0001} \Rightarrow \text{quotient} = 00010_{BIN}$$

Exercise 3.7:

```
 1          ;*********************************************************************
 2          ;*              Division of BIN numbers byte/byte            *
 3          ;*              differential method II                       *
 4          ;*********************************************************************
 5  000A             n EQU 10                      ;n=10 DEC
 6  0003             y EQU 3                       ;y=3 DEC
 7
 8  0000: 74 0A      MOV A,#n                      ;dividend
 9  0002: 75 F0 03   MOV B,#y                      ;divisor
10  0005: 12 00 0A   LCALL DIV_BIN8BY8DIFF
11                                                 ;result in A-quotient,
                                                    B-reminder
12  0008: 80 FE      STOP: SJMP STOP
13          ;----------------------------------------------------------------
14  000A:            DIV_BIN8BY8DIFF:
15  000A: AB F0      MOV R3,B
16  000C: BB 00 03   CJNE R3,#0,DIVIDE
17  000F: D2 D2      SETB OV
18  0011: 22         RET
```

```
19          0012:                DIVIDE:
20          0012: 79 08          MOV R1,#8
21          0014:                LOOP:
22          0014: C3             CLR C
23          0015: 33             RLC A                    ;<-divident
24          0016: CA             XCH A,R2                 ;<-reminder<-C
25          0017: 33             RLC A
26          0018: C3             CLR C
27          0019: 9B             SUBB A,R3                ;reminder-divisor
28          001A: 50 01          JNC NOT_LESS
29          001C: 2B             ADD A,R3
30          001D:                NOT_LESS:
31          001D: CA             XCH A,R2
32          001E: B3             CPL C
33          001F: C8             XCH A,R0
34          0020: 33             RLC A                    ;<-quotient
35          0021: C8             XCH A,R0
36          0022: D9 F0          DJNZ R1,LOOP
37          0024: E8             MOV A,R0
38          0025: 8A F0          MOV B,R2
39          0027: 22             RET
40          ;--- end of file ---
```

Exercise 3.8*: No solution is provided.

Exercise 3.9:

a.

$$
\begin{array}{r}
10010110_{P-BCD} \\
+ \quad 00010101_{P-BCD} \\
\hline
10101011_{P-BCD} \\
+ \quad 00000110 \\
\hline
10110001_{P-BCD} \\
+ \quad 01100000 \\
\hline
1 \quad 00010001_{P-BCD}
\end{array}
$$

b.

$$
\begin{array}{r}
10000100_{P-BCD} \\
+ \quad 01110011_{P-BCD} \\
\hline
11110111_{P-BCD} \\
+ \quad 01100000_{P-BCD} \\
\hline
1 \quad 01010111_{P-BCD}
\end{array}
$$

Exercise 3.10:

a.

$$
\begin{array}{r}
10010010_{P-BCD} \\
- \quad 10000111_{P-BCD} \\
\hline
00001011_{P-BCD} \\
- \quad 00000110 \\
\hline
00000101_{P-BCD}
\end{array}
$$

b.

$$
\begin{array}{r}
01100001_{P-BCD} \\
- \quad 00100101_{P-BCD} \\
\hline
00111100_{P-BCD} \\
- \quad 00000110 \\
\hline
00110110_{P-BCD}
\end{array}
$$

Exercise 3.11*: No solution is provided.

Exercise 3.12*: No solution is provided.

Exercise 3.13:

a.

$$
\begin{array}{r}
00001001\ \ 00000100_{UP-BCD} \\
+\ \ 00000101\ \ 00000010_{UP-BCD} \\
\hline
00001110\ \ 00000110_{UP-BCD} \\
+\ \ 11110110\ \ 00000000 \\
\hline
1\ \ 00000100\ \ 00000110_{UP-BCD}
\end{array}
$$

b.

$$
\begin{array}{r}
00001001\ \ 00000111_{UP-BCD} \\
+\ \ 00000101\ \ 00001000_{UP-BCD} \\
\hline
00001110\ \ 00001111_{UP-BCD} \\
+\ \ 11110110\ \ 11110110 \\
\hline
1\ \ 00000101\ \ 00000101_{UP-BCD}
\end{array}
$$

Exercise 3.14:

a.

$$
\begin{array}{r}
00001001\ \ 00000001_{UP-BCD} \\
-\ \ 00000110\ \ 00000011_{UP-BCD} \\
\hline
00000010\ \ 11111110_{UP-BCD} \\
-\ \ 00000000\ \ 11110110 \\
\hline
0\ \ 00000010\ \ 00001000_{UP-BCD}
\end{array}
$$

b.

$$
\begin{array}{r}
00000101\ \ 00000110_{UP-BCD} \\
-\ \ 00000100\ \ 00000010_{UP-BCD} \\
\hline
00000001\ \ 00000100_{UP-BCD}
\end{array}
$$

Exercise 3.15*: No solution is provided.

Exercise 3.16*: No solution is provided.

Exercise 3.17:

a.

$$
\begin{array}{r}
00110011\ \ 00110010_{ASCII} \\
\downarrow \\
00000011\ \ 00000010 \\
+\ \ 00110111\ \ 00110001_{ASCII} \\
\hline
00111010\ \ 00110101 \\
+\ \ 11110110\ \ 00000000 \\
\hline
1\ \ 00110000\ \ 00110101_{ASCII}
\end{array}
$$

b.

$$
\begin{array}{r}
00110101\ \ 00110110_{ASCII} \\
\downarrow \\
00000101\ \ 00000110 \\
+\ \ 00110111\ \ 00110001_{ASCII} \\
\hline
00111100\ \ 00111001 \\
+\ \ 11110110\ \ 00000000 \\
\hline
1\ \ 00110010\ \ 00111001_{ASCII}
\end{array}
$$

Exercise 3.18:

a.

$$00111001 \ 00110010_{ASCII}$$
$$- \ 00110111 \ 00110011_{ASCII}$$
$$\downarrow$$
$$00111001 \ 00110010_{ASCII}$$
$$- \ 00000111 \ 00000011$$
$$\overline{00110010 \ 00101111}$$
$$- \ 00000000 \ 11110110$$
$$\overline{00110001 \ 00111001_{ASCII}}$$

b.

$$1 \ \ 00110101 \ 00110110_{ASCII}$$
$$- \ 00110111 \ 00110010_{ASCII}$$
$$\downarrow$$
$$00110101 \ 00110110_{ASCII}$$
$$- \ 00000111 \ 00000010$$
$$\overline{1 \ 00101110 \ 00110100}$$
$$- 11110110 \ 00000000$$
$$\overline{00111000 \ 00110100_{ASCII}}$$

Exercise 3.19*: No solution is provided.

Exercise 3.20*: No solution is provided.

Exercise 3.21:

a.
$$\underline{1}100_{SM} \qquad \underline{1}0100_{SM}$$
$$+ \ \underline{1}111_{SM} \Rightarrow + \ \underline{1}0111_{SM}$$
$$\overline{?01 \ 1_{SM}} \qquad \overline{\underline{1}1101 \ 1_{SM}}$$

b.
$$\underline{0}100_{SM} \qquad \underline{0}0100_{SM}$$
$$+ \ \underline{0}111_{SM} \Rightarrow + \ \underline{0}0111_{SM}$$
$$\overline{?01 \ 1_{SM}} \qquad \overline{\underline{0}1011_{SM}}$$

c.
$$\underline{1}100_{SM}$$
$$+ \ \underline{0}111_{SM} \qquad |A|<|B| \qquad 0111$$
$$\overline{?} \qquad \xrightarrow{\hspace{1cm}} \ \frac{- \ 0100}{\underline{0} \ 0011_{SM}}$$

d.
$$\underline{0}100_{SM}$$
$$+ \ \underline{1}111_{SM} \qquad |A|<|B| \qquad 0111$$
$$\overline{?} \qquad \xrightarrow{\hspace{1cm}} \ \frac{- \ 0100}{\underline{1} \ 0011_{SM}}$$

In cases (a) and (b), there was a carry-over to the sign bit, so the numbers had to be written on five bits.

Exercise 3.22:

a.
$$\underline{1}100_{SM}$$
$$- \ \underline{1}111_{SM} \qquad |A|<|B| \qquad 0111$$
$$\overline{?} \qquad \xrightarrow{\hspace{1cm}} \ \frac{- \ 0100}{\underline{0} \ 0011_{SM}}$$

b.
$$\underline{0}100_{SM}$$
$$- \ \underline{0}111_{SM} \qquad |A|<|B| \qquad 0111$$
$$\overline{?} \qquad \xrightarrow{\hspace{1cm}} \ \frac{- \ 0100}{\underline{1} \ 0011_{SM}}$$

c.

$$\begin{array}{r} \underline{1}\,100_{SM} \\ -\ \underline{0}111_{SM} \\ \hline ?01\ _{SM} \end{array} \Rightarrow \begin{array}{r} \underline{1}\,100_{SM} \\ +\ \underline{0}111_{SM} \\ \hline ?01\ _{SM} \end{array} \Rightarrow \begin{array}{r} \underline{1}\,0100_{SM} \\ +\ \underline{0}0111_{SM} \\ \hline \underline{1}\,1101_{SM} \end{array}$$

d.

$$\begin{array}{r} \underline{0}100_{SM} \\ -\ \underline{1}\,11_{SM} \\ \hline ?01\ _{SM} \end{array} \Rightarrow \begin{array}{r} \underline{0}100_{SM} \\ +\ \underline{1}\,11_{SM} \\ \hline ?01\ _{SM} \end{array} \Rightarrow \begin{array}{r} \underline{0}0100_{SM} \\ +\ \underline{1}\,0111_{SM} \\ \hline \underline{0}101\ _{SM} \end{array}$$

In cases (c) and (d), there was a carry-over to the sign bit, so the numbers had to be written on five bits.

Exercise 3.23:

a.

$$\begin{array}{r} \scriptstyle 11 \\ 1\,001_{2's} \\ +\ 111\,1_{2's} \\ \hline \mathit{1}\ 1000_{2's} \end{array}$$

b.

$$\begin{array}{r} \scriptstyle 11 \\ 1\,011_{2's} \\ +\ 0110_{2's} \\ \hline \mathit{1}\ 0001_{2's} \end{array}$$

Exercise 3.24:

a.

$$\begin{array}{r} \scriptstyle 00 \\ 1\,101_{2's} \\ -\ 001\ 1_{2's} \\ \hline \emptyset\ 1010_{2's} \end{array}$$

b.

$$\begin{array}{r} \scriptstyle 10 \\ 0101_{2's} \\ -\ 1100_{2's} \\ \hline \mathit{1}\ 000 1_{2's} \end{array} \Rightarrow \begin{array}{r} \scriptstyle 11 \\ 00101_{2's} \\ -\ 11100_{2's} \\ \hline \mathit{1}\ 01001_{2's} \end{array}$$

In case (b), there was a carry-over to the sign bit position, so the numbers had to be written using five bits.

Exercise 3.25:

a)

$$\begin{array}{r} 1\ \ 1\ \ 0.\ \ 1 = A_{2's} \\ *\ \ \emptyset\ \ 0.\ \ 1\ \ 1 = \tilde{B} \\ \hline \underline{1}\ \ \underline{1}\ \ 1\ \ 1\ \ 0\ \ 1 \\ \underline{1}\ \ \underline{1}\ \ 1\ \ 0\ \ 1 \\ +\ 0\ \ 0\ \ 0\ \ 0 \\ \hline 1\ \ 1\ \ 1\ \ 0\ \ 1\ \ 1\ \ 1 \quad \text{pseudoproduct} \\ -\ 0\ \ 0\ \ 0\ \ 0\ \ 0\ \ 0\ \ 0 \quad \text{correction} \\ \hline 1\ \ 1\ \ 1\ \ 0.\ \ 1\ \ 1\ \ 1 \quad _{2's} \end{array}$$

b)

$$\begin{array}{r} 1\ \ 0\ \ 0\ \ 1 = A_{2's} \\ *\ \ \cancel{1}\ \ 1\ \ 0\ \ 0 = \tilde{B} \\ \hline \underline{0}\ \ \underline{0}\ \ 0\ \ 0\ \ 0\ \ 0 \\ \underline{0}\ \ \underline{0}\ \ 0\ \ 0\ \ 0 \\ +\ 1\ \ 0\ \ 0\ \ 1 \\ \hline \underline{1}\ \ 1\ \ 0\ \ 0\ \ 1\ \ 0\ \ 0 \quad \text{pseudoproduct} \\ -\ 1\ \ 0\ \ 0\ \ 1\ \ 0\ \ 0\ \ 0 \quad \text{correction} \\ \hline 0\ \ 0\ \ 1\ \ 1\ \ 1\ \ 0\ \ 0 \quad _{2's} \end{array}$$

Exercise 3.26:

a)

```
        |  |  |  |  |  |  0  |₂'ₛ
     *  0  0  0  0  0  0  1  |₂'ₛ
        |  |  |  |  |  |  0  |
     |  |  |  |  |  |  0  |
  0  0  0  0  0  0  0  0
              ⋮
 +     0  ⋯  0
  0  0  |  0  |  |  |  |  0  |  |  |₂'ₛ
```

b)

```
                    0  0  0  0. |  |₂'ₛ
                 *  |  |  |  |  0. 0₂'ₛ
                    0  0  0  0  0  0
              0  0  0  0  0  0
           0  0  0  0  |  |
        0  0  0  0  |  |
     0  0  0  0  |  |
  +  0  0  0  0  |  |
     Ø  Ø  Ø  /  Ø  |  |  0. |  0  0₂'ₛ
```

Exercise 3.27:

a.

```
        | | 0 |₂'ₛ
     *  0 0 | |₂'ₛ
        0000001{1̲0̲}
     −  | | 0 |
        00| |00|{1̲0̲}→
        000| |00{| 1̲}→
   ⎧ 0000| |0{0 1̲}
 + ⎨
   ⎩ | | 0 |
        | | 0 | | | 0{0 1̲}→
        | | | 0 | | |{00̲}→
        ⋯⋯⋯⋯⋯⋯⋯⋯⋯
   =    | | | | 0 | | |₂'ₛ
```

b.

```
        0 | |₂'ₛ
     *  | 0 0₂'ₛ
        000| 0{0̲0̲}→
        0000|{0̲0̲}→
        ⎧ 00000{1̲0̲}
     − ⎨
        ⎩ 0| |
        | 0| 00{1̲0̲}→
        ⋯⋯⋯⋯⋯⋯⋯⋯⋯
     =  | | 0| 00₂'ₛ
```

Exercise 3.28*: No solution is provided.

Exercise 3.29: No solution is provided.

Exercise 4.1: $M_{10} = -5.28_{DEC}$ i $E_{10} = -3_{DEC}$, hence:

$$E_2 = 1 + \frac{\ln|M_{10}| + E_{10}\cdot\ln 10}{\ln 2} = 1 + \frac{\ln|-5,28| + (-3)\cdot\ln 10}{\ln 2} = -6.565 \approx -7$$

$$= \underline{1}\,| | |_{SM}$$

$$M_2 = M_{10} \cdot \exp(E_{10} \cdot \ln 10 - E_2 \cdot \ln 2) = -5.28 \cdot \exp(-3 \cdot \ln 10 - (-7) \cdot \ln 2)$$
$$= -0.67584_{DEC}$$

0.67584·2
1.35168·2
0.70336·2
1.40672·2
0.81344·2
1.62688·2
1.25376·2
0.50752·2
1.01504·2
...
$0.67584_{DEC} \rightarrow \approx 0.10101101_{BIN}$

The result is $\underline{1}.10101101_{SM} \cdot 2^{\underline{1}111}{}_{SM}$

Checking: $-0.67584 \cdot 2^{-7} = -0.00528_{DEC}$.

Exercise 4.2: $M_2 = \underline{0}.0101_{SM} = +5/16_{DEC}$ and $E_2 = \underline{0}10_{SM} = +2_{DEC}$, hence:

$$E_{10} = 1 + \frac{\ln|M_2| + E_2 \cdot \ln 2}{\ln 10} = 1 + \frac{\ln|+0.3125| + (+2) \cdot \ln 2}{\ln 10} = 1.097 \approx 1$$

$$M_{10} = M_2 \cdot \exp(E_2 \cdot \ln 2 - E_{10} \cdot \ln 10) = +0.3125 \cdot \exp(2 \cdot \ln 2 - 1 \cdot \ln 10) = +0.125$$

The result is $+0.125 \cdot 10^{+1} = +1.25_{DEC}$

Exercise 5.1:

$$p = 2, \quad E_X = +10_{SM}, \quad M_X = -0.101_{SM}, \quad E_Y = -01_{SM}, \quad M_Y = +0.111_{SM}$$

$$|E_X - E_Y| = |+2-(-1)| = |+3| = +3, \quad E_X > E_Y$$

$M'_X = M_X$, $M'_Y = M_Y \cdot 2^{-3} = \underline{0}.0001111_{SM}$

Because X and Y are different signs and $|M'_X| > |M'_Y|$ (see addition rules for SM numbers – Table 3.1):

$$M_Z = M'_X + M'_Y = -(|M'_X| - |M'_Y|)$$
$$= -0.100100_{SM}, \quad E_Z = \max(+2, \ -1) = +2$$

The result is $Z = \underline{1}.1001001 \cdot 2^{\underline{0}10}{}_{SM} = -2.28125_{DEC}$.

It takes 8 bits to express the exact mantissa value of the result!

Exercise 5.2*: $p = 2$, $E_X = +00_{SM}$, $M_X = -0.101_{SM}$, $E_Y = +10_{SM}$, $M_Y = -0.0110_{SM}$

$|E_X - E_Y| = |+0 - (+2)| = |-2| = +2$, $E_X < E_Y$

$$M'_X = M_X \cdot 2^{-2} = -0.00101_{SM}, \quad M'_Y = M_Y$$

Because X and Y are negative and $|M'_X| < |M'_Y|$ (see subtraction rules for SM numbers – Table 3.2):

$$M_Z = M'_X - M'_Y = +(|M'_Y| - |M'_X|)$$
$$= +0.1001_{SM}, \quad E_Z = \max(+0, +2) = +2$$

The result is $Z = \underline{0}.10011 \cdot 2^{\underline{0}10}{}_{SM} = +2\frac{3}{8}{}_{DEC}$

Exercise 5.3:

$$p = 2, \quad E_X = +01_{SM}, \quad M_X = -0.0100_{SM}, \quad E_Y = -10_{SM}, \quad M_Y = -0.0010_{SM}$$

$$M_Z = M_X \cdot M_Y = +|M_X| \cdot |M_Y| = \underline{0}.00001000_{SM}, \quad E_Z = E_X + E_Y = +1 + (-2)$$
$$= -1_{DEC}$$

Because $|M_Z| < 2^{-4}$ the normalization is needed:

$$M_{Z\_norm} = M_Z \cdot 2^4 = \underline{0}.1000_{SM}, \quad E_{Znorm} = E_Z - 4 = -1 - 4 = -5 = \underline{1}101_{SM}$$

The result is $Z = \underline{0}.1000 \cdot 2^{\underline{1101}}{}_{SM} = +\frac{1}{64}{}_{DEC}$.

It takes 4 bits to express the exact exponent value of the result!

Exercise 5.4:

$$p = 2, \quad E_X = +10_{SM}, \quad M_X = +0.1101_{SM}, \quad E_Y = -10_{SM}, \quad M_Y = -0.1110_{SM}$$

Because $|M_X| < |M_Y|$ the denormalization of X number is not required and:

$$M'_X = M_X, \quad E'_X = E_X,$$

$$M_Z = M_X/M_Y = -|M'_X|/|M_Y| = \underline{1}.111011(011)...{}_{SM}, \quad E_Z = E'_X - E_Y$$
$$= +2 - (-2) = +4_{DEC}$$

The result is $Z \approx \underline{1}.111011(011)... \cdot 2^{\underline{0}100}{}_{SM} \approx -14.857_{DEC}$

It takes an infinite number of bits to express the exact mantissa value of the result!

Exercise 6.1:

a. $A = \textbf{round}(A',k)$
   - for $k = 3$ $A = 3.142 \cdot 10^{-2}$
     Absolute rounding error: $\Delta A = A - A' = -0.0000041$
     We have $|-0.0000041| < 0.000005$
     and $0.5ulp = 0.5 \cdot 10^{-k} \cdot 10^{E} = 0.5 \cdot 10^{-3} \cdot 10^{-2} = 0.000005$
   - for $k = 4$ $A = 3.1416 \cdot 10^{-2}$
     Absolute rounding error: $\Delta A = A - A' = -0.0000001$
     We have $|-0.0000001| < 0.0000005$
     and $0.5ulp = 0.5 \cdot 10^{-k} \cdot 10^{E} = 0.5 \cdot 10^{-4} \cdot 10^{-2} = 0.0000005$

b. $A = \textbf{truncate}(A',k)$
   - for $k = 3$ $A = 3.141 \cdot 10^{-2}$
     Absolute truncating error: $\Delta A = A - A' = -0.0000059$
     We have $-0.00001 < -0.0000059 < 0$
     and $ulp = 10^{-k} \cdot 10^{E} = 10^{-3} \cdot 10^{-2} = 0.00001$
   - for $k = 4$ $A = 3.1415 \cdot 10^{-2}$
     Absolute truncating error: $\Delta A = A - A' = -0.00009$
     We have $-0.0000001 < -0.00009 < 0$
     and $ulp = 10^{-k} \cdot 10^{E} = 10^{-4} \cdot 10^{-2} = 0.000001$

Exercise 6.2: $A' = 543$, $\Delta A = 2$, $B' = 398$, $\Delta B = 3$ hence: $\delta A = 2/543 \approx 0.37\%$, $\delta B = 3/398 \approx 0.75\%$

$$\delta_{A+B} = \frac{\Delta A + \Delta B}{A' + B'} = \frac{2 + 3}{543 + 398} = \frac{5}{941} \approx 0.5\% < 0.37\% + 0.75\%$$

$$\delta_{A-B} = \frac{\Delta A - \Delta B}{A' - B'} = \frac{2 - 3}{543 - 398} \approx -0.7\%$$

$$\delta_{A \cdot B} \approx \delta_A + \delta_B = 1.12\%$$

$$\delta_{A/B} \approx \delta_A - \delta_B = -0.38\%$$

Exercise 6.3: $A' = \frac{3}{16}$ $B' = \frac{5}{8}$ $A' = 0.001_{BIN}$ $B = 0.101_{DEC}$, hence:

$$\Delta A = A - A' = \frac{1}{8} - \frac{3}{16} = -\frac{1}{16} \quad \Delta B = B - B' = \frac{5}{8} - \frac{5}{8} = 0$$

ad a)

$$\Delta A + \Delta B = -\frac{1}{16} + 0 = -\frac{1}{16}$$

Checking: $A + B = 0.001_{BIN} + 0.101_{BIN} = 0.110_{BIN} = \frac{6}{8}$ $A' + B' = \frac{3}{16} + \frac{5}{8} = \frac{13}{16}$

$$\Delta_{A+B} = (A + B) - (A' + B') = \frac{6}{8} - \frac{13}{16} = -\frac{1}{16}$$

ad b)

$$\Delta A - \Delta B = -\frac{1}{16} - 0 = -\frac{1}{16}$$

Checking: $A - B = 0.001_{BIN} - 0.101_{BIN} = -\frac{4}{8}$ $A' - B' = \frac{3}{16} - \frac{5}{8} = -\frac{7}{16}$

$$\Delta_{A-B} = (A - B) - (A' - B') = -\frac{4}{8} - \left(-\frac{7}{16}\right) = -\frac{1}{16}$$

ad c)

$$B' \cdot \Delta A + A' \cdot \Delta B = \frac{5}{8} \cdot \left(-\frac{1}{16} \right) + \frac{3}{8} \cdot 0 = -\frac{5}{128}$$

Checking: $A \cdot B = 0.001_{BIN} \cdot 0.101_{BIN} = 0.000101_{BIN} = \frac{5}{64}$ $A' \cdot B' = \frac{3}{16} \cdot \frac{5}{8} = \frac{15}{128}$

$$\Delta_{A \cdot B} = (A \cdot B) - (A' \cdot B') = \frac{5}{64} - \frac{15}{128} = -\frac{5}{128}$$

ad d)

$$\frac{\Delta A}{B'} - \frac{A' \cdot \Delta B}{B' \cdot B'} = -\frac{1}{16} \cdot \frac{8}{5} - \frac{3}{16} \cdot 0 \cdot \frac{8}{5} \cdot \frac{8}{5} = -\frac{1}{10}$$

Checking: $A/B = 0.001_{BIN} / 0.101_{BIN} = \frac{1}{8} \cdot \frac{8}{5} = \frac{1}{5}$ $A'/B' = \frac{3}{16} \cdot \frac{8}{5} = \frac{3}{10}$

$$\Delta_{A/B} = (A/B) - (A'/B') = \frac{1}{5} - \frac{3}{10} = -\frac{1}{10}$$

Index

Printed in the United States
by Baker & Taylor Publisher Services